XR Development with Unity

A beginner's guide to creating virtual, augmented, and mixed reality experiences using Unity

Anna Braun

Raffael Rizzo

‹packt›

BIRMINGHAM—MUMBAI

XR Development with Unity

Group Product Manager: Rohit Rajkumar

Publishing Product Manager: Kaustubh Manglurkar

Book Project Manager: Sonam Pandey

Senior Editors: Divya Anne Selvaraj and Mudita Sonar

Technical Editor: Simran Ali

Copy Editor: Safis Editing

Proofreader: Safis Editing

Indexer: Hemangini Bari

Production Designer: Alishon Mendonca

DevRel Marketing Coordinator: Nivedita Pandey

First published: November 2023

Production reference: 1021123

Published by Packt Publishing Ltd.

Grosvenor House

11 St Paul's Square

Birmingham

B3 1RB, UK

ISBN 978-1-80512-812-0

www.packtpub.com

Life is a precious gift, my dear,

Embraced with those we hold near.

With endless love, beyond what I can pay back,

Thank you, mom and dad, for the support I never lack.

My brothers, my heart's compass and guide,

Near or far, with me, they always reside.

With you, dear husband, each moment we share,

A bond of love, laughter, and support beyond compare.

To my loving aunts, uncles, and cousins so dear,

Philippe, Catherine, Noora, and Jussi, your warmth is ever near.

Childhood friends, through decades we steer,

Invaluable moments, held forever dear.

Never forget, the gift of life is so rare,

Spend it with love, and show the world you care.

To neighbors, animals, and nature, be kind,

Let's foster love and care in our human mind.

Let's ensure life on Earth does shine,

For all beings, through the threads of time.

Our vision, vast as the cosmic sea,

To spread love throughout eternity.

– Anna Braun

This book stands as a significant milestone in my journey, but it pales in comparison to the achievements represented by the wonderful souls who surround me. To my parents, who laid the foundation of my life with their unwavering love and support; to my cherished brother and sister, who have walked this journey alongside me; to my friends, each one a beacon of light and strength; to my dear grandmother, whose nurturing spirit continues to guide me like a second mother; and to my future wife, soon to be the mother of our children, who ceaselessly inspires me to evolve into the best version of myself – this work is a testament to the love, guidance, and resilience you've instilled in me. My heart holds nothing but boundless love for each and every one of you.

– Raffael Rizzo

Contributors

About the authors

Anna Braun is an experienced XR developer and has already worked in XR departments at esteemed institutions, such as Fraunhofer IGD and Deutsche Telekom AG. She holds a profound interest in crafting **Augmented Reality** (**AR**) applications, tailored for both smartphones and AR glasses. Beyond her master's degree in extended reality, Anna is a distinguished author in the technological domain and is a frequent speaker at academic conferences and events, sponsored by non-profit organizations such as the Mozilla Foundation. Moreover, she is a co-founder of a firm specializing in XR consultancy and development.

To those who've shaped my journey – my parents, who offered endless support; my two brothers, who served as my compass; and my dear husband, with whom every shared moment is a testament to unparalleled love, laughter, and support. To family and friends whose warmth remained constant through time's tide. May this work reflect our shared belief in the boundless capacity of love and the interconnectedness of all life.

Raffael Rizzo initiated his XR career by developing a VR training program for a soccer academy, aimed at assessing young athletes' reaction times. He has worked at esteemed institutions such as Deutsche Telekom AG and Fraunhofer IGD, amassing a wealth of experience in XR with a focus on Unity. His master's degree encompasses extended reality, computer vision, machine learning, and 3D visualization. Raffael has co-founded a company dedicated to XR consultations and development.

To the incredible tapestry of people who've shaped my life – my parents, siblings, friends, cherished grandmother, and future wife – this achievement pales beside the gift of having you all. My gratitude and love know no bounds.

About the reviewers

Darren Delorme is a senior Unity XR developer who has established himself as a seasoned professional in the world of augmented and virtual reality. In 2011, Darren created a map and app with the use of Vuforia, publishing and distributing 70,000 AR-enabled Tourism Maps in Tofino, British Columbia, Canada. In 2018, Darren joined Oculus Start, and in 2019, he worked as a Unity Live Help expert. Working at Arbor XR, Darren played a key role in developing a multi-platform input system for Quest, Pico, and Vive. He also designed the user interface for ArborXR home and created the 3D home environment for all their supported platforms. Darren recently joined the **Royal Astronomical Society of Canada** (**RASC**) and started a VR space project, slated for release in 2024, merging his passion for astronomy with his love for virtual reality.

Denis Pineda, known as `@dampheldev` on social media, is a visionary video game developer specializing in creating XR experiences. A self-taught 3D game artist since the age of 12, and passionate about learning new technologies, Denis proudly holds a bachelor's degree in game design, with a dedicated focus on developing Unity 3D and C# projects from scratch. His repertoire encompasses the creation of awe-inspiring 3D art, from modeling and texturing to animation and integration with real-time PC, mobile, and VR platforms. Denis has contributed his expertise to advergaming XR projects in Guatemala and Honduras. He has also worked as an indie game developer for his Rambutan Dog Games brand and as a sought-after freelancer.

Table of Contents

Part 2 – Interactive XR Applications with Custom Logic, Animations, Physics, Sound, and Visual Effects

3

VR Development in Unity   37

4

AR Development in Unity   65

5

Building Interactive VR Experiences   99

6

Building Interactive AR Experiences   125

7

Adding Sound and Visual Effects   153

Part 3 – Advanced XR Techniques: Hand-Tracking, Gaze-Tracking, and Multiplayer Capabilities

8

Building Advanced XR Techniques   181

9

Best Practices and Future Trends in XR Development   219

Preface

Hi there, and welcome to the exciting and ever-evolving world of Extended Reality (XR) development! If you're feeling a mix of excitement and nervousness about diving into XR development in Unity for the first time, you're not alone.

But don't worry – the primary goal of this book is to make XR development accessible to everyone, regardless of prior experience. We'll use simple language, provide vivid examples, and guide you through every step, from making objects grabbable to writing scripts in C# and incorporating hand-tracking into your scenes. You won't even need access to a VR headset or AR-compatible smartphone to follow along; you can test most of the projects using simulators on your laptop or PC, ensuring accessibility for all.

In our own XR development journey, we've faced the frustration of spending endless hours trying to make a basic XR experience work. Now that we've gained extensive experience at notable companies such as Deutsche Telekom and academic institutions such as Fraunhofer IGD, we want to share our knowledge with you.

This book is not only a complete guide that will take you from being a novice in XR development or Unity to reaching an intermediate level in creating interactive XR applications for any domain. Its structure is also designed to help you easily access XR-related techniques to build outstanding XR applications, so you won't need to constantly search online for answers, as we once did. Our goal is not to narrow your XR development expertise to specific use cases but to equip you with a wide array of tools so that you can bring any XR project to life, without limitations.

We're thrilled to start on this journey with you!

Who this book is for

If you're a student, professional, or just curious about venturing into VR, MR, or AR, this book is tailored for you. Whether you're familiar with interactive media or just beginning, we will guide you through building XR applications in Unity with ease. No Unity experience? No problem. We cover the essentials, equipping you to develop interactive VR, MR, and AR projects.

What this book covers

Chapter 1, Introduction to XR and Unity, serves as a general introduction to the topic of XR development in Unity. This chapter explains which approaches exist to bring VR, MR, and AR to life. Furthermore, Unity's role in XR development is introduced. The main goal of this chapter is to provide a good baseline to start learning how to build XR applications.

Chapter 2, The Unity Editor and Scene Creation, is aimed at those unfamiliar with Unity. It explains how to install Unity Hub and the Unity Editor and provides a step-by-step guide on how to create a basic scene in Unity. This chapter introduces fundamental concepts, from lighting and rendering to importing assets from the Unity Asset Store, equipping you with everything you need to know about the Unity Engine before diving into XR development.

Chapter 3, VR Development in Unity, presents the capabilities and components of the XR Interaction Toolkit and how to add it to a VR scene. After exploring the different types of interactions, from grabbing to climbing, this chapter explains how to test and deploy VR scenes to VR headsets or simulators, such as the XR Device Simulator.

Chapter 4, AR Development in Unity, explains how to create AR experiences in Unity using AR Foundation, ARKit, and ARCore. After building your first, simple AR application, this chapter focuses on testing this application directly on a PC, using XR simulation, and deploying it to Android and iOS devices.

Chapter 5, Building Interactive VR Experiences, explains how to add interactivity to a scene via animations, buttons events, or the programming language C#. Although this chapter teaches intermediate-level concepts, it is beginner-friendly and doesn't require any preexisting knowledge of C#.

Chapter 6, Building Interactive AR Experiences, details how to create an interactive AR application, via touch controls and UI elements. Using the programming language C#, this chapter showcases how to build an AR experience that aligns with state-of-the-art design patterns and the core components of commercial AR applications.

Chapter 7, Adding Sound and Visual Effects, covers how to make XR scenes more immersive and realistic by mimicking physical phenomena from real life. This chapter explains the foundation of sound theory and particle behavior and highlights how both of these physical phenomena can be simulated in an XR environment with Unity. Through a hands-on project, this chapter explains how to add audio sources, audio mixers, and a particle system to an XR scene and how to fine-tune their properties, making them as realistic as possible.

Chapter 8, Building Advanced XR Techniques, introduces advanced XR techniques, elevating the overall user experience and immersion of any XR application. Specifically, this chapter details how to add hand-tracking or gaze-tracking support to VR applications via the XR Interaction Toolkit and explains how to build multiplayer experiences, where users can see each other via simple avatars and hand animations.

Chapter 9, Best Practices and Future Trends in XR Development, explores the current and future trends in XR technology. It provides insights into XR research and XR companies of various industries and introduces best practices along the entire XR development life cycle. To conclude this book, this chapter introduces additional toolkits and plugins for XR development. These resources are ideal for further exploration in future XR projects beyond what was covered in this book.

To get the most out of this book

If you don't yet have Unity installed on your PC or laptop, don't worry – we will guide you through this process in *Chapter 2*.

Software/hardware covered in the book	Operating system requirements
Unity version 2021.3.4 or later	Windows, macOS, or Linux

As you venture into creating immersive virtual, augmented, and mixed reality experiences, it's crucial to set the foundation right. This starts by cloning the project repository using Git LFS. Even if you plan to build all projects from the ground up, it's highly recommended to clone the entire repository. There will undoubtedly be sections where cross-referencing your progress with the actual solution will prove beneficial.

> **Attention**
> Do not simply download the repository in a ZIP format. Doing so will most likely result in errors upon opening. The correct approach is to clone it using Git LFS.

When working on virtual, augmented, and mixed reality projects in Unity, it's common to handle large files – think 3D models, textures, audio clips, and more. Simply downloading a project might seem like a straightforward approach, but you may run into issues. The reason is rooted in how GitHub (and many other platforms) handle large files.

Let's illustrate this with a simple analogy. Imagine you have a vast library of books. Instead of storing all these books at your home, which would take up enormous space, you get a reference card for each book. Whenever you want to read a particular book, you present the card at the library, and they provide the book for you.

Git **LFS**, which stands for Git **Large File Storage**, operates on a similar principle. Instead of saving bulky files directly within the repository, Git LFS maintains a tiny reference or "pointer." The actual hefty files are stored elsewhere. Consequently, when you directly download a project without employing LFS, you only obtain these pointers, not the genuine files. To secure the entire content, you must clone the project with Git LFS.

The installation process for Git LFS and cloning our repository is very straightforward. Here's a step-by-step guide:

1. **Install Git**: If you have Windows, download Git from the Git for Windows website (`https://gitforwindows.org/`), and follow the installation instructions. For macOS, you can either download Git from the official website (`https://git-scm.com/download/mac`) or simply use `brew install git` in your Terminal if you have Homebrew. For Linux, use your package manager, such as `sudo apt-get install git` for Debian-based distributions.

2. **Setting up Git**: Open your Terminal or Command Prompt. Configure your username with `git config --global user.name "Your Name"`. Configure your email with `git config --global user.email "youremail@example.com"`.

3. **Install Git LFS**: Visit the official Git LFS website (`https://git-lfs.com/`) and follow their installation instructions.

4. **Initialize Git LFS**: In your Terminal or Command Prompt, type `git lfs install`. This sets up Git LFS for your user account.

5. **Sign up for GitHub**: If you haven't already, create a free account on GitHub (`https://git-lfs.com/`).

6. **Clone the Repository**: Navigate to the directory where you want to clone the project. Open your Terminal. Use the `git clone https://github.com/PacktPublishing/XR-Development-with-Unity.git` command.

That's it! You're now ready to work with the Git repository and Git LFS to manage large files.

If you are using the digital version of this book, we advise you to type the code yourself or access the code from the book's GitHub repository (a link is available in the next section). Doing so will help you avoid any potential errors related to the copying and pasting of code.

Downloading the example code files

You can download the example code files for this book from GitHub at `https://github.com/PacktPublishing/XR-Development-with-Unity`. If there's an update to the code, it will be updated in the GitHub repository. Do remember to clone using Git LFS, as described previously, instead of simply downloading the repository in a ZIP format. Doing so will ensure that there are no errors upon opening.

We also have other code bundles from our rich catalog of books and videos available at `https://github.com/PacktPublishing/`. Check them out!

Additional Resources

The demo videos of the projects created in this book are available in the GitHub repository: `https://github.com/PacktPublishing/XR-Development-with-Unity/tree/main/Additional%20Resources`

Conventions used

There are a number of text conventions used throughout this book.

`Code in text`: Indicates code words in text, database table names, folder names, filenames, file extensions, pathnames, dummy URLs, user input, and Twitter handles. Here is an example: "Our final step is to override the `Update()` function."

A block of code is set as follows:

```
private void Update()
{
float scaleValue = slider.value;
bus.transform.localScale = new Vector3(scaleValue, scaleValue,
scaleValue);
}
```

Any command-line input or output is written as follows:

```
$ mkdir css
$ cd css
```

Bold: Indicates a new term, an important word, or words that you see on screen. For instance, words in menus or dialog boxes appear in **bold**. Here is an example: "Within Unity Hub, navigate to the **Installs** tab and hit the **Add** button, to add a new Unity Editor version."

> **Tips or important notes**
> Appear like this.

Get in touch

Feedback from our readers is always welcome.

General feedback: If you have questions about any aspect of this book, email us at `customercare@packtpub.com` and mention the book title in the subject of your message.

Errata: Although we have taken every care to ensure the accuracy of our content, mistakes do happen. If you have found a mistake in this book, we would be grateful if you would report this to us. Please visit www.packtpub.com/support/errata and fill in the form.

Piracy: If you come across any illegal copies of our works in any form on the internet, we would be grateful if you would provide us with the location address or website name. Please contact us at copyright@packt.com with a link to the material.

If you are interested in becoming an author: If there is a topic that you have expertise in and you are interested in either writing or contributing to a book, please visit authors.packtpub.com.

Share Your Thoughts

Once you've read *XR Development with Unity*, we'd love to hear your thoughts! Scan the QR code below to go straight to the Amazon review page for this book and share your feedback.

https://packt.link/r/1-805-12812-4

Your review is important to us and the tech community and will help us make sure we're delivering excellent quality content.

Download a free PDF copy of this book

Thanks for purchasing this book!

Do you like to read on the go but are unable to carry your print books everywhere?

Is your eBook purchase not compatible with the device of your choice?

Don't worry, now with every Packt book you get a DRM-free PDF version of that book at no cost.

Read anywhere, any place, on any device. Search, copy, and paste code from your favorite technical books directly into your application.

The perks don't stop there, you can get exclusive access to discounts, newsletters, and great free content in your inbox daily

Follow these simple steps to get the benefits:

1. Scan the QR code or visit the link below

https://packt.link/free-ebook/9781805128120

2. Submit your proof of purchase
3. That's it! We'll send your free PDF and other benefits to your email directly

Part 1 –
Understanding the Basics
of XR and Unity

Welcome to the first part of this book, where we'll provide you with essential information about Extended Reality (XR) technologies and the Unity Engine. This part serves as a foundation for your journey in XR application development in the subsequent parts. Here, we will guide you through the process of installing Unity Hub and the Unity Editor, and we'll introduce you to important concepts such as scene creation, asset downloading from Unity Asset Store, the various types of light sources available in Unity, and explanations of supported rendering types. No prior knowledge of Unity or XR technologies is required to follow along with this part. Instead, this part will equip you with all the necessary knowledge to start your hands-on XR development adventure.

This part comprises the following chapters:

- *Chapter 1, Introduction to XR and Unity*
- *Chapter 2, The Unity Editor and Scene Creation*

1

Introduction to XR and Unity

In this chapter, we will explore the extraordinary potential of **XR development** and how **Unity**, the leading game engine in the field, can help us unlock it. Through this journey, you will gain a deep understanding of the various forms of **Extended Reality** (**XR**) and their capabilities, paving the way for you to create immersive and unforgettable experiences. You will discover how Unity plays a crucial role in blending the realms of the physical and digital worlds worldwide. Let's dive into this fascinating world and discover the limitless possibilities that await us.

In this chapter, our discussion will encompass the following topics:

* Understanding XR and its different forms (AR, MR, and VR)

* How did Unity evolve as a platform for XR development?

Understanding XR and its different forms (VR, AR, and MR)

XR is an all-encompassing term that includes a range of technologies designed to enhance our senses. Whether it's providing additional information about the physical world or creating entirely new, simulated worlds for us to explore, XR comprises **Virtual Reality** (**VR**), **Augmented Reality** (**AR**), and **Mixed Reality** (**MR**) technologies. You may be someone who has already put on multiple **XR headsets**, lost yourself in immersive **VR games** such as *Beat Saber*, or even have a **VR headset** of your own. If so, you already have a pretty good idea of what XR is. You may know that while VR implies a complete immersion experience that shuts out the physical world, AR and MR combine the real and virtual worlds with each other. However, can you pinpoint the differences between AR and MR?

What is the difference between AR and MR?

Let's explore this through an example – imagine you're wearing a *Meta Quest 3* headset. This headset comes equipped with cameras that allow passthrough capabilities – a feature enabling you to view your real-world surroundings while also being immersed in the virtual environment. When these cameras are active, you're essentially in an MR experience. For instance, as you stand in your living room, you might spot an ancient vase on your coffee table. In actuality, your coffee table is real, but the vase is a virtual object seamlessly integrated through MR. The true magic of MR lies in its ability to intertwine the digital and physical worlds so intimately that it becomes challenging to differentiate between them. When the passthrough is disabled, however, the headset returns to a purely VR experience, detaching you from the physical world's view.

In contrast, AR enhances our experience of the real world by projecting digital information onto it. When wearing **AR glasses**, such as *HoloLens* or *Magic Leap*, there's a clear delineation between the real and virtual elements in your view. For instance, imagine wearing AR glasses and seeing interactive labels hovering next to objects in your room, such as a floating note reminding you to water a plant or arrows directing you to a misplaced item.

Additionally, there are mobile AR devices, where the augmentation is experienced through a smartphone or tablet screen. These devices provide a windowed experience of the augmented world – for example, games such as *Pokémon Go*, where you use your smartphone's camera to spot virtual creatures in the real world, or an app that lets you visualize how a piece of furniture looks in your room before purchasing.

AR strives to enhance our current reality, but MR pushes the boundaries of imagination to envision a world where real and virtual objects merge seamlessly, while VR creates a completely alternative reality altogether. Many find it challenging to differentiate between MR headsets and AR glasses, but the display serves as a good indicator. AR glasses utilize transparent displays instead of cameras, granting them direct access to real-world lighting for seamless integration of digital overlays. In contrast, MR headsets allow users to view the environment through additional sensors and cameras, but they do not provide direct access to sunlight. It's like a spectrum of reality – at one end, we have the familiar and the real, subtly enhanced with digital elements; at the other end, we have entirely invented worlds, with limitless potential for exploration and adventure. And somewhere in between, we have the magical space where the real and virtual merge, seamlessly blending the best of both worlds. Can you guess which idea dates back the earliest – VR, AR, or MR? And how did XR enter the picture?

How did AR, VR, MR, and XR evolve?

Although VR certainly is the most advanced of these technologies to date, it was the author of *The Wonderful Wizard of Oz*, L. Frank Baum, who predicted AR in his 1901 novel *The Master Key: An Electrical Fairy Tale and the Optimism of Its Devotees*. Nevertheless, it was not until half a century had passed that the first aspirations of meandering through virtual realms and augmenting our reality truly took flight.

In 1962, Morton Heilig introduced the *Sensorama* – a device that plays with more senses than modern VR headsets. In one of the Sensorama's experiences, you were teleported into a mythical jungle, and could hear the calls of tropical birds, smell the moist soil, and feel the mild breeze on your face as you wandered on a thin trail. In another experience, you rode a quad through a desert, inhaling the smell of hot rocks through your nose, feeling the wind hitting your cheeks from different angles as you made a turn, and listening to the symphony of the wind flowing down the dunes and the quad's rusty motor. The Sensorama consisted of a stereoscopic color display, fans to simulate wind, odor emitters creating different smells with chemicals, a movable chair, and a stereo-sound system (`https://www.youtube.com/watch?v=vSINEBZNCks`).

The emergence of XR technology was a rocky and unpredictable journey. Following the Sensorama's sensory immersion, which painted a splendid vision of blended realities within reach, the 1990s witnessed a VR terrain strewn with the debris of ill-fated contraptions, such as *Sega VR* and *Nintendo Virtual Boy*. They failed to ignite the consumer's imagination. As a result, the momentum of **Head-Mounted Displays** (**HMDs**) began to wane.

This is why researchers at Indiana University invented a completely different XR technology called the **Computer Automatic Virtual Environment** (**CAVE**). This technology enables viewers standing inside a cube and wearing 3D glasses to see any 3D graphics projected onto the walls, ceiling, and floor as if they're floating in space. Users wander through the scenes, observing the projected objects from any angle. CAVEs can usually host up to 10 people, all of whom might be standing in front of a virtual replica of Michelangelo's David statue, viewing the masterpiece from any angle, plunging into the rollercoaster of evoked emotions, and capturing every slight edge with their eyes. As CAVEs enabled high-resolution XR experiences at a time when heavy HMDs with low-quality resolution were the only alternative, they seemed to be the most promising way to experience XR for a few years. So, which factors led to the widespread adoption of HMDs to experience virtual words, while the seemingly superior CAVEs failed to gain significant traction?

Installing a CAVE costs a pile of money, often hundreds of thousands of dollars, paired with high monthly maintenance and usage costs. Setting up a CAVE can be tiring. You need to properly adjust the projectors, add appropriate lighting and air conditioning, and coordinate a network of computers to ensure the projectors seamlessly synchronize with one another. Most importantly, CAVEs fall short when it comes to multiple individual experiences or immersive storylines that demand interactive engagement between the users and the application: `https://sky-real.com/news/the-vr-cave-halfway-between-reality-and-virtuality/` and `https://web.archive.org/web/20070109083006/http://inkido.indiana.edu/a100/handouts/cave_out.html`.

After a bleak winter in XR development, marked by the dashed excitement over CAVEs and a slew of failed HMDs, a glimmer of hope emerged in the early 2000s. In 2005, the world witnessed the birth of the first-ever **AR app** called *AR Tennis*. This game could be played by two people, armed only with a *Nokia* phone and a sense of adventure. Using the phone's camera, a marker was tracked, and voilà – a tennis court sprang to life on the screen. Both participants could interact with a virtual ball, court, and opponent, all simulated within the device. The mobile phone itself functions as the racquet, allowing for an immersive gameplay experience (`https://www.imgawards.com/games/ar-tennis/`).

In 2012, *Palmer Luckey*, often regarded as the father of modern VR for his pioneering work, marked another milestone in XR development with the introduction of a crowdfunded VR headset known as the *Oculus Rift*. Luckey founded *Oculus VR*, the project's parent company, which was eventually acquired by *Facebook* in 2014 (`https://history-computer.com/oculus-history/`). The following year, *Microsoft* unveiled its groundbreaking *HoloLens* **AR headset**, which seamlessly blended high-definition holograms with the real world (`https://news.microsoft.com/en-au/2016/10/12/microsoft-announces-global-expansion-for-hololens/`). In 2016, *Niantic* released *Pokémon Go*, a game that would go on to capture the hearts of millions and usher in a new era of **AR games**. Like wildfire, the game's popularity spread across the globe, igniting a new wave of excitement and enthusiasm for AR technology.

In 2017, *IKEA* introduced its famous AR app, allowing customers to transcend the limits of imagination and see how IKEA's furniture would look in their homes, without ever leaving the comfort of their own couch.

Today, we have arrived at a pivotal moment in XR history, where the technology has never been more thrilling, and the possibilities never more endless. From its humble beginnings as a niche industry, XR is now poised to explode into every aspect of our lives, like a supernova of innovation and progress. In 2021, Mark Zuckerberg unveiled his bold vision for the future – a world where immersive, practical **XR devices** are not just a novelty but also an essential tool to enhance our lives and make our everyday joys even more extraordinary. The year 2022 saw the launch of the *Meta Quest Pro*, marking the first time *Meta* dipped its toe into releasing **MR headsets**, after covering a global market share of 81% with their VR headsets, such as the *Meta Quest 2* (`https://www.counterpointresearch.com/global-xr-ar-vr-headsets-market-share/`).

However, even as XR technology becomes more advanced and sophisticated, there are still some who believe in making it accessible to all. The founders of the **Open Source Community for Augmented Reality** (**OpenAR**), a group dedicated to democratizing AR, have created DIY Open AR Glasses, a budget-friendly project that empowers anyone to build their own **AR glasses** with just €20 (`https://sites.uef.fi/openar/` and `https://openar.fi/`). It is beautiful to witness how technology can be used to bring people together and make the world a better place.

How did Unity evolve as a platform for XR development?

Unity Technologies, now a billion-dollar titan in the realms of gaming and XR, emerged from humble origins in a cozy Copenhagen flat in 2004. The company's journey to dominating the market with its *Unity game engine* has been equally fraught with uncertainty and turbulence, like that of XR technologies.

In 2002, a Danish graphics student sought aid on a digital forum, where a Berliner high schooler with backend experience joined him. Serendipity united their game studio dreams. With only a father's financial cushion and café wages, they created their first game in 2005 – an alien, trapped in a life-support sphere, navigating Earth. Players tilted the world, and the alien rolled, gathering gems and evading CIA capture, utilizing wall-sticking and jumping tricks. The game's difficulty overshadowed its cutting-edge lighting, and it found little success. Recognizing their talent for building tools and prototypes, the duo refocused their energy on crafting a game engine for the Mac community, dubbing it Unity – a symbol of collaboration and compatibility (`https://techcrunch.com/2019/10/17/how-unity-built-the-worlds-most-popular-game-engine/`).

Today, Unity's fame in the realm of mobile game creation is legendary, with good cause. As the architect behind nearly half the globe's games, it stands as the model for countless developers to aspire to. Yet, Unity's reach extends far beyond just gaming. It's increasingly used for 3D design and simulations across other industries such as film, automotive, and architecture. In fact, Unity is used to create a stunning 60% of all AR and VR experiences worldwide (`https://www.dailydot.com/debug/unity-deempind-ai/`).

In the world of high-end filmmaking, VR and the Unity Game Engine are the hottest ingredients to create a cinematic masterpiece. Disney's Oscar winner *The Jungle Book*, Spielberg's *Ready Player One*, and Oscar winner *Blade Runner 2049* all owe their stunning visuals to these breakthrough technologies (`https://unity.com/madewith/virtual-cinematography`). *The Lion King*, a technical wonder in the realm of photo-realistic animation, is no exception. From every bone and breath to each muscle and whisker, every aspect of the African lions was carefully observed in nature and simulated in Unity to mimic real-life lions as closely as possible, while retaining the distinguished features of the original animated movie (`https://ai.umich.edu/blog-posts/how-disneys-the-lion-king-became-a-pioneer-in-the-use-of-virtual-reality/`).

Beyond the realm of filmmaking, XR development in Unity has the potential to transform the very way we communicate and interact with each other. Through the use of immersive virtual environments and cutting-edge **Artificial Intelligence** (**AI**) and emotion systems, XR has the power to create non-linear narratives, personalized experiences, and dynamic interactions that blur the lines between reality and fantasy. From education and training to gaming and social media, the possibilities for XR are endless, and Unity is at the forefront of this exciting new frontier.

XR development in Unity enabled experiences unlike anything the world had ever seen before, such as the award-winning VR game *Bonfire*, where you're stranded on a desolate planet with only a flickering bonfire. Your survival depends on building a rapport with an extraterrestrial who speaks no known language and relies solely on your non-verbal cues and gestures. The result is magical, non-verbal communication, accompanied by a completely organic, non-linear storyline, driven by sophisticated AI and emotion systems that adapt to your every move in real time.

Unity's XR development not only transforms gaming but also revolutionizes healthcare, as exemplified by *VirtaMed's* innovative applications that reshape the way surgeons acquire surgical skills. The **MR simulator** mimics every action of a real surgical operation, from the subtlest movements to complications such as unexpected bleeding and bile leakage. Every trainee's move is recorded with millimeter accuracy, providing valuable feedback. Simulated training accelerates the learning curve for aspiring surgeons, allowing them to enter the operating room with greater confidence and skill.

Summary

This chapter introduced you to the different facets of XR, namely AR, MR, and VR. You learned how AR strives to enhance our current reality, how MR pushes the boundaries of imagination to envision a world where real and virtual objects merge seamlessly, and how VR creates a completely alternative reality altogether. You now understand how the early dreams of delving into completely virtual or augmented realities transformed into bold actions and how XR development was and still is far from a straightforward path. You learned about Unity, from its humble beginnings to its worldwide dominance in game development, simulations, and XR development, looking at some state-of-the-art use cases. The following chapter will familiarize you with the Unity Hub and Editor. You will learn how to import assets, create materials, and acquire all the other skills you need before you can start developing your first XR experience.

2

The Unity Editor and Scene Creation

In this chapter, we'll lay the groundwork for your Unity journey. You'll familiarize yourself with the Unity Editor, create a basic scene, and explore essential lighting aspects. We'll cover installing Unity, navigating the Editor, working with **GameObjects**, importing assets, and experimenting with various lighting settings. By the end, you'll have a solid foundation to delve deeper into Unity and create increasingly complex and captivating scenes.

We'll cover the following topics as we proceed:

- Setting up the Unity development environment
- Getting to know the Unity Editor and its interface
- Understanding GameObjects and components
- Creating a basic scene in Unity and adding objects

Technical requirements

Before diving into the Unity Editor, it is important to ensure that your system meets the minimum requirements to run Unity. To successfully complete the exercises in this chapter, you will require a personal computer that has *Unity 2021.3* LTS or a more recent version installed. To ensure your hardware meets the requirements, you can cross-check it on the Unity website (`https://docs.unity3d.com/Manual/system-requirements.html`).

Setting up the Unity development environment

First things first, let's get Unity up and running on your development machine. Throughout this book, we'll be harnessing the power of the Unity 3D game engine to create inspiring projects. Unity is an incredibly potent, cross-platform 3D development environment, complete with an intuitive and visually appealing editor.

If you have not yet installed Unity on your computer, we will guide you through the process. Following the installation, we'll proceed to create our initial scene. Let's begin the setup and exploration of Unity.

Installing the Unity Hub

Over the course of this book, the **Unity Hub** will become your trusty command center for managing different Unity projects, Unity Editor versions, and modules. To initiate the installation process of the Unity Hub, follow these steps:

1. Head over to the official Unity website (`https://unity3d.com/get-unity/download`) and navigate to the latest version of the Unity Hub.

2. Follow the onscreen instructions to install Unity Hub.

3. With the Unity Hub installed, open it up and sign in using your Unity account. If you're new to Unity, create an account to join the ranks of fellow creators.

Without having the Unity Editor installed, the Unity Hub is just as powerful as a CD player without a CD. The next section covers how you can install the Unity Editor within the Unity Hub.

Installing the Unity Editor

The **Unity Editor** is where the magic happens—a workspace for designing, building, and testing your game projects. To install it, follow these steps:

1. Within the Unity Hub, navigate to the **Installs** tab and hit the **Add** button to add a new Unity Editor version.

2. Opt for the latest LTS version of the Unity Editor and click **Next** to kick off the installation process.

3. During installation, don't forget to include the necessary platforms and modules tailored to your specific needs. Add the *Windows/Mac/Linux Build Support*, depending on the operating system of your PC. Likewise, select the *Android* or *iOS Build Support*, depending on the nature of your smartphone, so that you can follow along with the AR tutorials in this book. Lastly, if you are using a VR headset that runs on Android, such as the *Quest 2* or *Quest Pro*, be sure to add the *Android Build Support* module along with its sub-modules: *OpenJDK* and *Android SDK & NDK Tools*.

As we have the Unity Editor installed, it is time to create a project.

Loading a sample scene as a new Unity project

After installing the Unity Hub and the Unity Editor, it's time to create a new Unity project. For the sake of simplicity, we will first use a sample scene. A sample scene in Unity is a pre-built scene created by Unity to show developers how various functions and techniques can be implemented. Unity offers a variety of sample scenes, ranging from simple 2D games to complex 3D environments, that can be used as a starting point for your own projects. The sample scenes can be downloaded via the Asset Store or directly from the Unity Hub. Here's how to do that directly from the Unity Hub.

Figure 2.1 shows a project's creation within the Unity Hub.

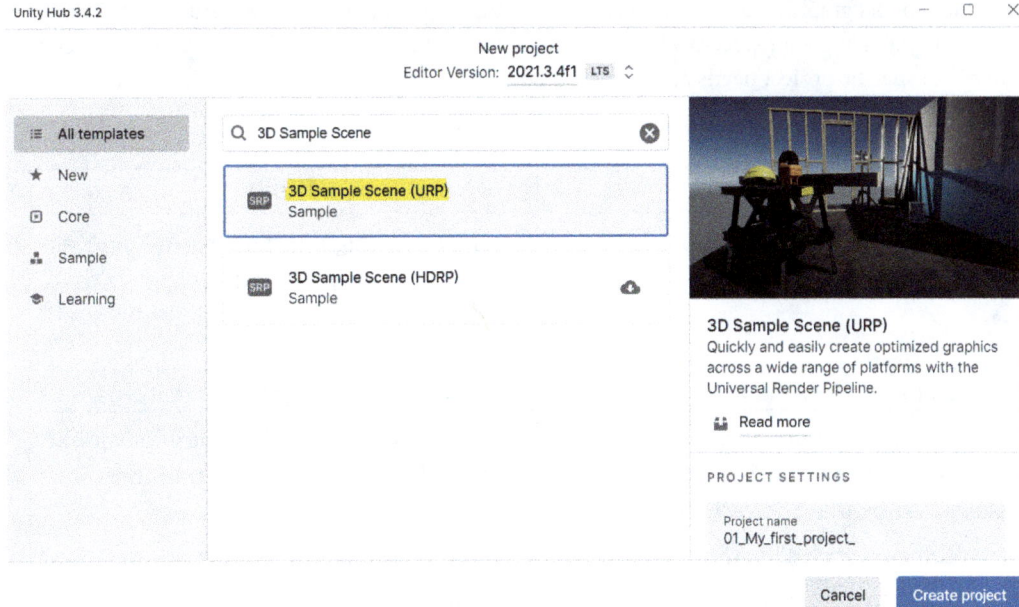

Figure 2.1 – How to create a project with a sample scene in the Unity Hub

To load a sample scene as a new project, just go through the following steps:

1. Open the Unity Hub and go to the **Projects** tab.

2. Click on the **New** button to create a new project.

3. Choose **3D Sample Scene (URP)**, give your project a name, and select a location to save it. For this project, we've opted for **3D Sample Scene (URP)** as it provides a preconfigured environment showcasing the capabilities of the **Universal Render Pipeline** (**URP**), ideal for those new to Unity or seeking a reference. While the standard **3D (URP)** template is typically favored by developers for its clean slate, allowing for a customized setup, they often enhance these projects by importing additional packages or assets via the Package Manager or Unity Asset Store.

4. Click on the **Create** button to create the project.

Now that we've got that out of the way, let's look at how to choose the right render pipeline.

Choosing the right render pipeline

Upon examining the sample scene options, you might have noticed the choice between the URP and the **High Definition Render Pipeline** (**HDRP**). But which one is better, and what exactly is a render pipeline? In essence, a render pipeline is a sequence of steps and processes that dictate how the engine renders graphics. It transforms 3D assets, lighting, and other scene components into the final 2D image gracing your screen. While URP and HDRP share some low-level tasks, each pipeline is tailored to specific project needs and target platforms.

Table 2.1 shows how URP and HDRP stack up against one another:

URP	HDRP
Designed for a wide range of platforms, including mobile devices, consoles, and PCs.	Designed for high-end platforms, such as PCs and consoles, targeting projects that demand high-quality graphics and visual fidelity.
Focus on performance and scalability.	Focus on advanced graphics features.
Provides a good balance between graphics quality and performance, making it a suitable choice for many game projects, including VR.	Is generally not recommended for mobile or other low-end platforms due to its high performance requirements.

Table 2.1 – A comparison between URP and HDRP

A comfortable VR experience demands high frame rates, typically exceeding 90 FPS. URP emphasizes performance, ensuring smooth frame rates across various VR devices, including standalone VR headsets, PC-based VR, and mobile VR. As the market sees a growing number of standalone VR headsets, URP proves invaluable for its adaptability and ease of project optimization.

While URP may not boast the visual prowess of HDRP, it strikes a balance between graphic quality and performance, making it suitable for the majority of VR projects where performance is crucial.

Unity's versatility extends beyond HDRP and URP pipelines, allowing the creation of custom **Scriptable Render Pipeline** (**SRP**) pipelines for experienced graphics programmers. However, developing a custom SRP pipeline requires deep knowledge of 3D graphics programming, rendering pipelines, and C# language proficiency. For those lacking these skills, HDRP and URP offer an optimal balance between flexibility and ease of use.

Given its many advantages, URP emerges as the go-to choice for most VR endeavors. Throughout this book, we'll focus exclusively on URP. However, HDRP remains a worthy contender for those pursuing high-end PC VR experiences.

With Unity installed, the render pipeline selected, and your project at your fingertips, it's time to acquaint ourselves with the Unity Editor.

Getting to know the Unity Editor and its interface

If you're new to Unity, the editor's interface can be a bit overwhelming at first. But don't worry—we'll guide you through the Unity Editor and show you how to navigate its various menus and panels. Experienced users can also benefit from staying current with the latest best practices and techniques, as designing for VR presents unique challenges that may require a different approach than traditional game development.

Exploring the Unity interface

Upon launching a new Unity project, you'll be greeted by the Unity Editor. This multifaceted workspace is composed of several distinct windows known as panels.

Figure 2.2 shows the window layout for the sample scene project we just created.

Figure 2.2 – The window layout for the sample scene project

Figure 2.2 showcases a number of panels, namely: (*1*) **Scene** view, (*2*) **Game** view, (*3*) **Hierarchy**, (*4*) **Inspector**, (*5*) **Project**, and (*6*) **Console**.

Let's explore these essential panels that make up Unity's interface.

Imagine the **Scene view** (*1*) as your canvas, where you'll bring your game world to life, creating mesmerizing landscapes and placing your characters in fantastic environments. This panel is the heart of the game, where every object is placed and arranged to tell a compelling story. For example, you can select the **Safety Hat** GameObject within our sample scene and move, rotate, scale, or remove it.

The **Game view** (*2*) is the place where you can experience your game from your player's perspective. It provides a real-time preview of the gameplay, including the visual rendering and user interface elements you've implemented.

The **Hierarchy panel** (*3*) or **Scene Hierarchy window** is your game's blueprint, showcasing the organized list of every GameObject that makes up your game world. It's like an architectural plan that helps you navigate, manage, and visualize the relationships between game elements, ensuring a coherent and structured experience. Our sample scene demonstrates what a well-structured hierarchy looks like. All added GameObjects are subordinate to the **Example Assets** parent object. This includes the **Props** GameObject, which itself serves as a parent object for any GameObjector asset that decorates a scene and adds details and context to the game world—for example, Jigsaw, Hammer, Workbench, and so on. So, keep in mind that grouping related objects together under a parent GameObject makes it easier to manage and manipulate them as a single entity. Additionally, naming conventions can be used to make the hierarchy more readable and easier to understand.

The **Inspector panel** (*4*) or **Inspector window** is the control center for fine-tuning your game elements. It's where you can adjust every tiny detail of your GameObjects or assets, making sure your game world is precisely how you envisioned it. From position and scale to adding components and modifying scripts, the **Inspector** panel is your ticket to perfection. Let's select the **Safety Hat** object in the **Scene Hierarchy** window by navigating to **Example Asset | Props | Safety Hat**. The **Inspector** window shows all the defined components of the **Safety Hat** object. The **Transform** component is responsible for positioning, rotating, and scaling the object. You will find a gizmo representation of it in the Scene View, which allows us to transform objects directly there. Two other components that can be seen in the **Inspector** window are the **Mesh Filter** and **Mesh Renderer** components. The **Mesh Filter** and **Mesh Renderer** components provide a way to create and display 3D models in a scene. The **Mesh Filter** component defines the model's geometry, while the **Mesh Renderer** component applies visual properties such as materials and textures. Without these components, you wouldn't see the object in the Scene View.

The **Project** panel (*5*) holds all the building blocks of your game, from **textures** and models to sounds and scripts. It's like a library of resources, where every imported or created asset is at your fingertips, waiting to be used in your XR project.

Finally, the **Console** panel (*6*) is an essential tool, assisting in the identification and resolution of issues during development. It provides detailed logs, warnings, and error messages, allowing for efficient troubleshooting and ensuring the integrity of the game's performance.

You are now familiar with the default panels in Unity. Next, we will get to know the Unity Grid and Snap system, which is a game changer when building scenes.

> **Tip**
> You can keep the default layout, or customize your panels using the dropdown at the top right of the Unity Editor under **Layout**. We usually prefer the 2x3 layout, but for simplicity, we will use the default layout in this book.

Using the Grid and Snap system

The **Unity Grid and Snap system** helps align and place objects in a more organized manner in the game environment. It allows you to snap objects to a grid and also to other objects for easier placement and arrangement.

Figure 2.3 shows you where to find the system.

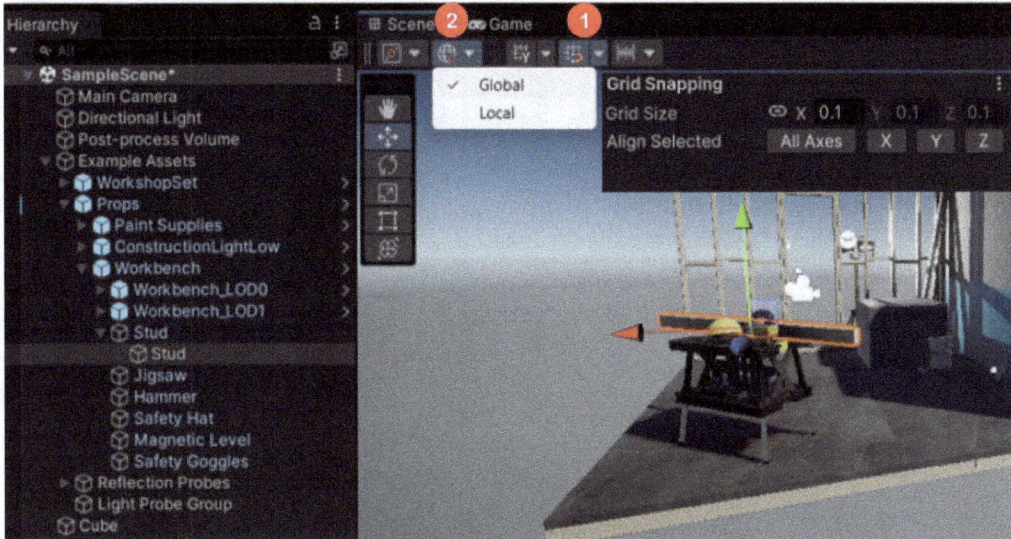

Figure 2.3 – How to use the Grid and Snap system

To make use of this system, you will need to turn on the **Grid Snapping** button, which is represented by an icon with a grid and magnet (*1*). Furthermore, make sure to activate the **Global** handles by selecting the **Global** icon (*2*) positioned adjacent to the **Grid Snapping** field.

The **Grid Size** field in the **Grid Snapping** dropdown in Unity refers to the size of the grid squares in the **Scene** view. By default, the **Grid Size** field is set to 1 for all three axes (**X**, **Y**, and **Z**), meaning each grid square has a width, height, and depth of one unit. This setting can be adjusted as needed to match the scale of the objects in your scene. The **Grid Size** field determines the increment at which objects will snap to the grid, so a larger grid size will result in coarser snapping, while a smaller grid size will result in finer snapping.

> **Your turn**
>
> 1. Select the same stud as in *Figure 2.3* and move it two units towards the **Y** direction. Activate **Grid Snapping**, and try different grid sizes to place the stud back on the workbench.
>
> 2. Try to become familiar with the navigation in the **Scene** view by moving, orbiting, and zooming through the scene. This documentation may help you: `https://docs.unity3d.com/510/Documentation/Manual/SceneViewNavigation.html`).
>
> 3. Pick another two GameObjectsand move, scale, and rotate them as you like.

Understanding GameObjects and components

GameObjects and **components** are essential building blocks of Unity projects, allowing developers to create interactive and dynamic content. Now, we'll provide a comprehensive overview of GameObjects and components and how they work together.

Understanding the default new scene

Unity empowers developers with the ability to create multiple scenes within the editor. This feature aids in managing complexity, enhancing performance, and fostering more modular and reusable game projects. To create a new default scene alongside our sample scene, access the menu at the uppermost section of the editor and choose **File | New Scene | Standard (URP)**. Save it in your `Scenes` folder with a name of your choosing. After following these steps, you will be presented with a new Unity scene containing a **Main Camera** GameObject and a **Directional Light** GameObject. Both GameObjects are located in the **Scene Hierarchy** window and displayed in the **Scene** window, which features an infinite reference ground plane grid. In Unity, GameObjects serve as the foundational building blocks of a game scene. They represent visible or interactive elements such as characters, props, light sources, or cameras. Each GameObject possesses a **Transform** component that defines its position, rotation, and scale and can be augmented with additional components for **physics**, **collision detection**, or **game logic**.

Let's create our first GameObject. Keep in mind that while Unity is primarily a game engine, it does offer some basic modeling capabilities. You can create simple 3D models using Unity's built-in 3D object **primitives** or import more intricate models fashioned in other 3D modeling software. Third-party plugins are also available to further bolster Unity's modeling prowess. However, it's important to remember that Unity is no match for dedicated 3D modeling software such as *Blender*, *3ds Max*, or *Maya*, and it's better suited for creating and manipulating models for use in game development.

We'll start by creating a plane primitive to act as our floor. To do so, right-click in the **Scene Hierarchy** window and select **3D Objects | Plane**. Notice how your newly created plane becomes visible in the **Scene** view. Navigate to the plane's **Inspector** window, and observe that it is not situated at the origin (0,0,0) but rather at a different location. By default, newly created objects are parented to the active object in the **Scene Hierarchy** window when created. If another object were active, the new object's position would be relative to that object's transform, possibly landing it somewhere unexpected. To reposition our plane at the origin, right-click on **Transform** in the **Inspector** window and select **Reset**. Next, right-click on the plane in the **Scene Hierarchy** window and rename it `Ground`.

In the `Ground` object's **Inspector** window, you'll notice that besides the **Mesh Filter** and **Mesh Renderer** components responsible for the GameObject's visual appearance, there's also a **Mesh Collider** component. Up next, you'll learn more about this and other types of colliders.

Adding a collider

Colliders are components that define the shape of an object and are used to determine physical interactions with other objects in the scene. There are several types of colliders available in Unity, each with its own specific use case, as follows:

- **Box collider**: This type of collider defines a rectangular shape, perfect for objects with a box-like shape.
- **Sphere collider**: This type of collider defines a spherical shape, great for objects with a round shape, such as a ball.
- **Capsule collider**: This type of collider is a combination of a cylinder and two spheres, useful for objects that are shaped like a capsule, such as a character.
- **Mesh collider**: This type of collider defines the shape of an object based on its mesh, allowing precise collision detection.
- **Wheel collider**: This type of collider is specifically designed for wheel-based objects, such as vehicles. It allows realistic wheel behavior and suspension.

By incorporating a collider component, you can define an object's interactions with other scene colliders, such as bouncing, stopping, or passing through. Additionally, colliders can trigger events, such as activating sound effects upon collision. Applying a collider to the ground is sensible, as it establishes environmental boundaries and enables lifelike interactions with other scene objects, preventing characters from falling through the ground, for instance.

Components are versatile tools that can be assigned to GameObjects, shaping their behavior, appearance, and functionality. Consider **Main Camera**, which is equipped with components such as **Transform** (defining position, rotation, and scale) and **Camera** (specifying settings such as **field of view** (**FOV**) and clipping planes). Similarly, the **Directional Light** GameObject features components such as **Transform** and **Light**, which determine the light's type, color, and intensity, among other properties.

Other examples of components include **Rigidbody** for incorporating physics-based behavior, **AudioSource** for playing sounds, and **Particle System** for crafting visual effects such as fire, smoke, or magical spells. Each component fulfills a unique function and significantly influences your game world and its interactions.

As we progress through this book, we will explore and utilize an array of essential components, starting with the next section, where we breathe life into our initial basic scene.

Creating a basic scene in Unity and adding objects

In this section, we will walk you through constructing the scene depicted in *Figure 2.4*, which integrates elements such as primitives, materials, prefabs, lighting, and imported assets.

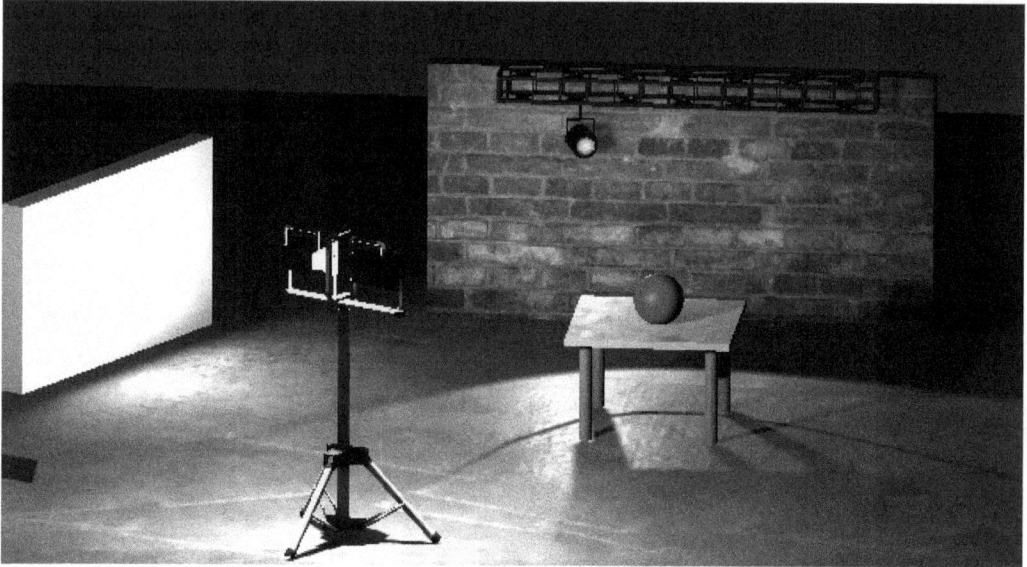

Figure 2.4 – The basic scene we are going to create

By engaging with this step-by-step tutorial, you will develop a strong knowledge of these core concepts while assembling an aesthetically striking scene. Our initial task involves crafting a table using primitive shapes.

Building a table with primitives

Primitives, as the name suggests, are simple geometric shapes that form the building blocks for more complex models in our scene. Imagine a chair taking shape as a cube forms the seat, two cylinders become the sturdy legs, and two more cylinders morph into supportive back legs. The backrest is crafted from a plane, its size and placement perfectly complementing the seat and legs.

A table for our scene can be crafted in a similar fashion—a cube takes center stage as the tabletop, while four cylinders rise up to serve as legs. Textures are applied, giving the tabletop and legs a lifelike appearance. The size and positioning of each primitive are adjusted, resulting in a table that looks like it was plucked from the real world.

To create the table, we recommend following the sequence illustrated in *Figure 2.5*.

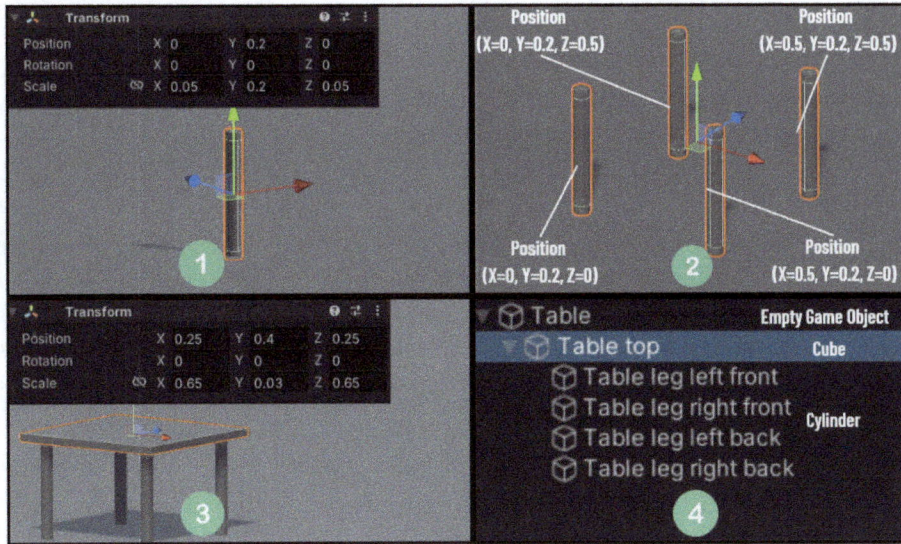

Figure 2.5 – How to use the Grid and Snap system

To build your table, go through the following steps:

1. Create a single table leg by introducing a cylinder to the scene. This can be achieved by right-clicking into the **Scene Hierarchy** window and selecting **GameObject | 3D Object | Cylinder**. Adjust the cylinder's scale to (0.05, 0.2, 0.05) and position at (0, 0.2, 0).

2. The table requires four identical legs. To achieve this, reproduce the original leg by copying (right-click on the object in the **Scene Hierarchy** window and click **Copy**) and pasting it three times. Fine-tune the positions of the duplicated legs based on the preceding reference image.

3. Next, add a **Cube** primitive to the scene that will serve as the tabletop. Rescale and position the cube so that it rests on the table legs. To achieve the scaling and positioning more simply, shift to **Wireframe** mode, as shown in *Figure 2.6*.

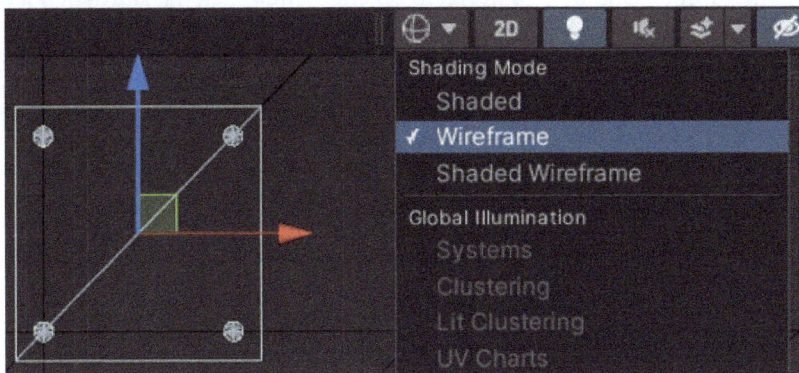

Figure 2.6 – How to use Wireframe mode

> **Tip**
>
> Our advice here is to initially scale just the **Y** value and place the cube on top of the legs using the Grid and Snap system. Subsequently, you can shift your view from the scene to the top and modify the **Shading** mode from **Shaded** to **Wireframe**. This change reveals the cylinders and the cube, simplifying the task of scaling the **X** and **Z** values and positioning the **Cube** object.

4. To organize the table components, create a nested parent-child object hierarchy. First, assign appropriate names to the objects by renaming the **Cube** object `Table top` and the cylinders as their respective table legs. Subsequently, select all the legs (*cmd/Ctrl* + left-click). Click and drag them onto the `Table top` object in the **Scene Hierarchy** window and release the mouse button. This action will make the table legs become child objects of the `Table top` object, establishing a parent-child relationship.

5. To further refine the table structure, create an empty GameObject by selecting **GameObject | Create Empty**, and rename it `Table`. Make the `Table top` object a child of the `Table` object. This results in a harmoniously structured table consisting of five interconnected primitives. With this arrangement, the table can be easily scaled and relocated without the need to individually select each GameObject.

To complete the scene, place a sphere primitive named **Sphere** on top of the `Table` object. You can use the Grid and Snap techniques discussed earlier for this. For a more refined look, feel free to scale the sphere to your liking. In the following section, we will enhance the appearance of both the `Table` and **Sphere** objects using materials.

Changing the appearance of the ground, table, and sphere

Up to this point, the scene consists of a ground and a table with a sphere placed on it, all displaying the standard gray color. Let's enhance this by creating a new red material for the sphere through the following steps:

1. Navigate to the `Materials` folder in the **Project** window (`Assets | Materials`).

2. Click on the + symbol and then **Material** to generate a new material and assign it the name `Red Material`.

3. Select the white square adjacent to the **Base Map** field in the **Inspector** window. This action will launch the **Color** window. Select the **#FF0000** hex code.

4. Now, you can apply the `Red Material` material by dragging and dropping it on top of the **Sphere** object.

Now, the sphere should appear in red. So far, we have used the simplest form of a material to apply a simple red color to a sphere geometry.

When creating materials for objects in a 3D game or application, it's often necessary to add additional details and depth to the surface of the object beyond what can be achieved with just a single texture. This is where maps such as base maps, metallic maps normal maps, height maps, and occlusion maps come in.

Let us get to know each of these maps, as follows:

- **Base Map**: A Base Map, also known as a diffuse map, is the main texture that defines the color and pattern of an object's surface. Think of it as the paint on a wall or the pattern on a piece of fabric. We can either assign an albedo color, which is a single color value that represents the overall diffuse color of an object's surface (as we did with the `Red Material` material) or we can assign a texture (such as a picture of a brick wall) that defines the color and pattern of an object's surface.

- **Metallic Map**: A Metallic Map defines the extent to which an object's surface should appear to be metallic, with areas of the map that are white appearing highly metallic, and areas that are black appearing non-metallic.

- **Normal Map**: A Normal Map is a type of texture that defines the surface normal vectors of an object, giving the illusion of additional depth and detail on the surface. Normal Maps are often used to create the appearance of bumps, dents, and other small details on an object's surface, without adding extra geometry.

- **Height Map**: A Height Map is a black-and-white image that defines the height of an object's surface. Height Maps are used to create the appearance of depth and relief on a surface, such as mountains and valleys on a terrain or the ridges and valleys of a fingerprint.

- **Occlusion Map**: An Occlusion Map is a texture that defines which parts of an object are occluded, or hidden, from view. Occlusion Maps are used to create the appearance of shadows and ambient occlusion, giving objects a more grounded, realistic look by making them appear to cast shadows and have depth.

To visualize these maps, imagine a piece of paper with a picture of a wall painted on it. The **Base Map** would be the paint on the wall, the **Normal Map** would add the appearance of bumps and dents to the surface, the **Height Map** would create the appearance of depth and relief, and the **Occlusion Map** would add the appearance of shadows. When these maps are combined, they can create a much more detailed and believable representation of a wall. You may have noticed that this example did not cover the **Metallic Map**. You do not have to use all of the maps for a material. The choice of which maps to use depends on the desired appearance of the material and the level of detail you want to achieve. For example, a simple material for a flat, single-colored object may only require a **Base Map** (or albedo color) to define its appearance. A more complex material, such as a metal object with intricate details, might require a **Base Map**, **Normal Map**, **Height Map**, and **Metallic Map** to achieve the desired look.

Creating maps can be a complex and time-consuming task, especially for intricate objects. To save time and effort, it's often best to utilize pre-made materials or models from marketplaces such as the **Unity Asset Store**, **ArtStation**, and **Sketchfab**. These sites offer a wealth of free content to choose from.

In our sample scene, we have a variety of pre-made materials available. Let's use a wooden material for the tabletop. Here's what you need to do:

1. Open the **Project** window and locate the `Materials` folder (`Assets | ExampleAssets | Materials`).

2. Search for the **OBS_Mat** material and drag it onto the top of the table.

Now, the tabletop should exhibit a wooden texture, as illustrated in *Figure 2.7*.

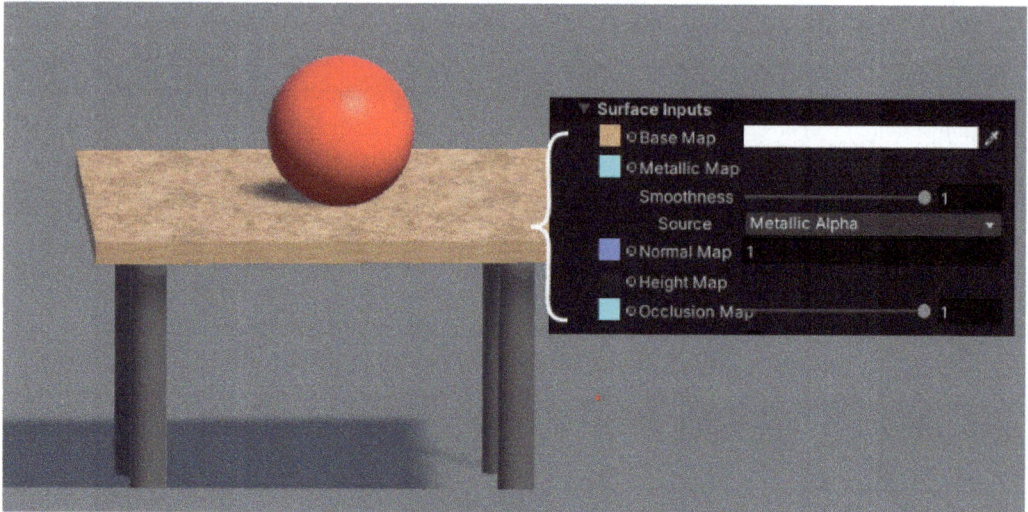

Figure 2.7 – The table top with its wooden texture

Upon examining the surface inputs in the **Inspector** window, the wooden material consists of a base map, metallic map, normal map, and occlusion map. To preview each map, *cmd/Ctrl* + left-click the corresponding squares next to the maps.

Let's also choose an appropriate material for the ground. In the **Project** window, locate the `Materials` folder (`Assets | ExampleAssets | Materials`) and search for the **Ground_Mat** material. Drag it onto the ground, and the surface should now appear more realistic.

Next, you will learn how to create a brick wall material in the scene.

Creating a brick wall material

In the previous sections, we explored the application of a red albedo color to a sphere and utilized more advanced materials for the ground and table. It's worth noting that materials play a crucial role in determining how a surface is rendered, encompassing not only the textures of an object's surface but also its interaction with light sources.

A texture is an image that is applied to a 3D model or a 2D surface to add detail or color. It can be thought of as skin that covers the surface of an object to give it a specific look or feel. Textures can be created within Unity or imported from external image files and then assigned to materials that are applied to GameObjects in a scene. This technique enhances the level of detail and realism in the rendered object.

In this section, we will delve into the process of creating and applying a more sophisticated material by developing a brick wall material and applying it to a wall constructed from a transformed cube. Follow these steps to get started:

1. Let's start by creating a wall by inserting a new cube into the scene, renaming it `Wall`, scaling it to (2, 1, 0.1), and positioning it at (0.2, 0.5, 1.8) behind the table.

2. Navigate to the `Assets` folder in the **Project** window and create a new folder by clicking on the + symbol and then **Folder**. Rename it `Textures`.

3. To give the `Wall` cube a brick-like appearance, download and utilize the `BrickWall.jpg` texture file that is provided in this book's GitHub repository, or find a suitable brick wall texture through an online search. Import the image into the newly created `Textures` folder. This can be accomplished by simply dragging the image from your local filesystem and dropping it into the corresponding Unity folder.

4. Staying in the **Project** window, progress to the `Materials` folder. Here, generate a new material and rename it `Brick_Mat`. You can create the material by right-clicking within the folder, selecting **Create**, and then choosing **Material**.

5. With the `Brick_Mat` material selected, navigate to the **Inspector** window. Look for the small *torus symbol* situated adjacent to the **Base Map** field and select it. Proceed to search for and select the imported **BrickWall** image.

6. The final step involves moving the `Brick_Mat` material from the **Project** window to the `Wall` cube present in the **Scene** window. This can be achieved via a simple drag and drop maneuver.

With these steps, your scene should now look similar to the one shown in *Figure 2.8*.

Figure 2.8 – The brick wall in the scene

You have now witnessed the prowess of materials in transforming a scene's realism. In the next section, we will learn about another important component of any Unity scene in more detail: prefabs.

Unpacking a prefab

In this section, we will learn how to convert a prefab from the sample scene into a regular GameObject in Unity. Prefabs are reusable, prefabricated objects that can be utilized multiple times within a scene, acting like building blocks for your environment. They are master copies of objects saved in separate files, allowing you to create multiple instances and streamline scene management.

To import a prefab, navigate to the **Project** window and select `Assets | ExampleAssets | Prefabs`. Drag and drop the **ConstructionLight_Low** object from the **Scene Hierarchy** window into the scene. In the **Scene Hierarchy** window, you'll notice a blue cube icon, indicating it is a prefab instance.

We'll now unpack the **ConstructionLight_Low** object to turn it into a regular GameObject. Right-click in the **Scene Hierarchy** window and select **Prefab | Unpack**. This way, modifications to the prefab won't affect the unpacked instance. Go ahead and position it in front of the brick wall.

It's important to remember that using prefabs in Unity is advantageous in various scenarios, such as the following:

- **Repeating GameObjects**: Save time and effort by creating an object once, converting it into a prefab, and duplicating it as needed

- **Customizing GameObjects**: Manage customizations more efficiently with a prefab containing all possible modifications, adjusting instances as required

- **Modular game design**: Facilitate a modular design by creating prefabs for individual-level components and assembling them into a complete level

Using prefabs in Unity helps save time, manage complex GameObjects, and support an adaptable, modular game design. Next, we'll discuss how to import assets from the Unity Asset Store.

Importing from the Unity Asset Store

Let's learn how to import and use models available on the Asset Store. The Unity Asset Store is a marketplace where users can find, purchase, and download a variety of assets, including 3D models, animations, audio, visual effects, and more, to use in their Unity projects.

Here's how to import a spotlight model from the Asset Store:

1. Open your preferred web browser and navigate to the Unity Asset Store at `https://assetstore.unity.com/`. Here, input `Spotlight and Structure` in the search bar. Alternatively, you can directly access the package page via `https://assetstore.unity.com/packages/3d/props/interior/spotlight-and-structure-141453`.

2. Once you have added the package to your assets, look for the **Open in Unity** button and click on it.

3. The Unity application should open with a prompt to import the package. Proceed with the **Import** operation.

4. Upon successful import, navigate to the package's directory in the **Project** window. The path should be `Assets | SpaceZeta_Spotlight | Assets`.

5. Drag the `Spotlight.fbx` and `SpotlightStructure1.fbx` files into the scene and unpack them (right-click, then select **Prefab | Unpack**).

If the objects appear magenta, this typically signifies that the associated materials are either absent or incompatible with the render pipeline currently in use. To fix this, select **Spotlight** and change the shader to **Universal Render Pipeline | Lit** in the **Inspector** window.

If your **Spotlight** model appears without any textures now, click on the small torus symbol situated adjacent to **Base**, **Metallic**, **Normal**, and **Emission Map** and assign the corresponding `Spotlight_basecolor.png`, `Spotlight_roughness.png`, `Spotlight_normal.png`, and `Spotlight_emission.png` files.

Now, your **Spotlight** and the **SpotlightStructure1** models should have the right texture as they share the same material. Rename them `BrickWallLight` and `BrickWallLightStructure` and transform them according to what is shown in *Figure 2.9*.

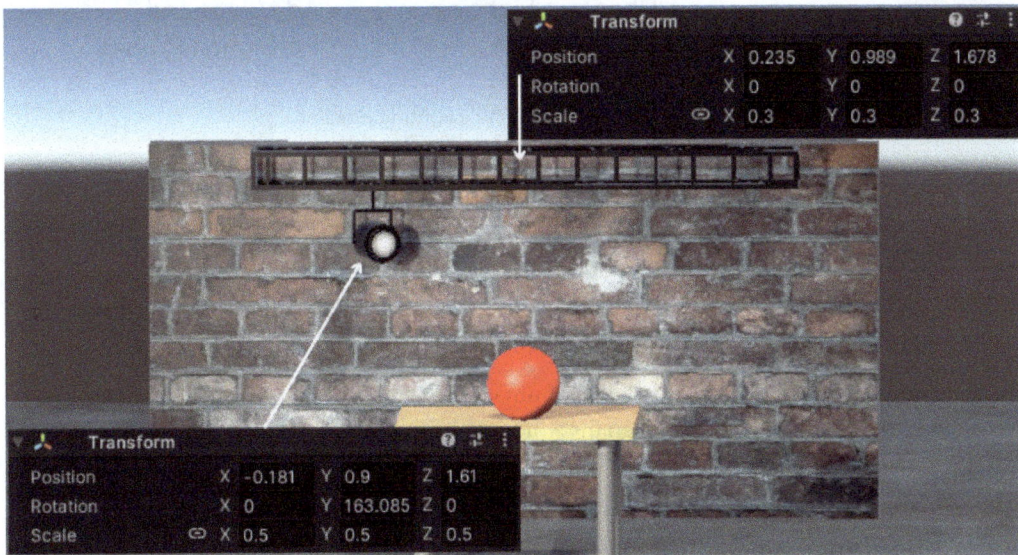

Figure 2.9 – The imported BrickWallLight model and its BrickWallLightStructure model

You can use the same procedure to find and import other 3D models, animations, audio, and visual effects that you need for your scene. In line with the `BrickWallLight` model that was imported just now, the next section deals with the most crucial aspects of lighting.

Understanding Unity's lighting pipeline

A crucial aspect to consider before implementing your project is planning the lighting. Here, we present a proven approach that has consistently delivered exceptional lighting in a wide array of XR projects over the years, closely aligning with Unity's best practice lighting pipeline (`https://docs.unity3d.com/Manual/BestPracticeLightingPipelines.html`). This process consists of three primary steps, as shown in *Figure 2.10*.

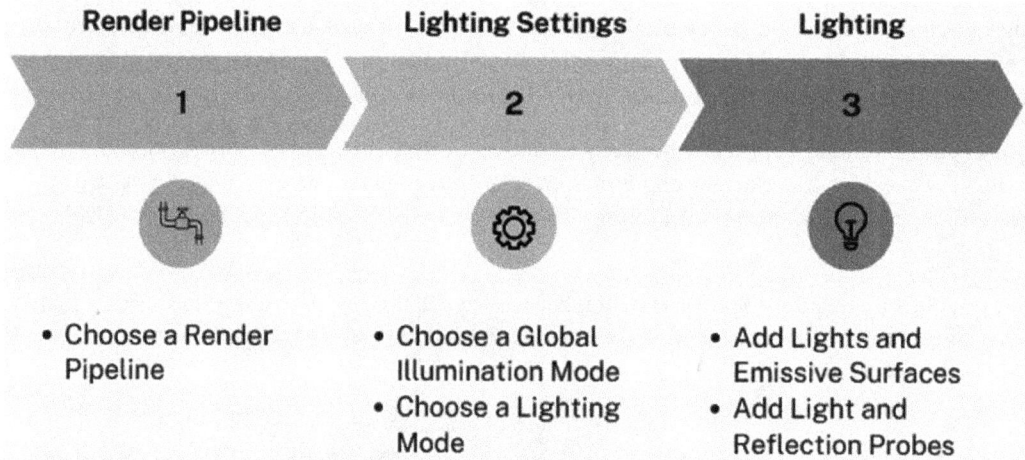

Render Pipeline	Lighting Settings	Lighting
1	2	3
• Choose a Render Pipeline	• Choose a Global Illumination Mode • Choose a Lighting Mode	• Add Lights and Emissive Surfaces • Add Light and Reflection Probes

Figure 2.10 – Unity's best practice lighting pipeline

The selection of an appropriate render pipeline for your project is your initial step of the overall lighting pipeline process. As noted in *Table 2.1*, we recommend using URP for XR projects, which establishes the groundwork for effective lighting.

The second step of the lighting pipeline is to determine how **indirect lighting** should be generated by selecting a suitable **Global Illumination (GI) system** and corresponding **lighting mode**.

Think about a room with a single window on a sunny day. The direct light is like the bright sunlight that's coming straight into the room through the window—it's the light you would feel directly on your skin if you stood in the path of the window. That's similar to what we mean by direct lighting in Unity.

Now, look at the corners of the room or the parts not directly in front of the window. They aren't in complete darkness, right? They're lit up, but not as brightly. That's because the sunlight coming in hits different surfaces—the floor, walls, or furniture—and bounces off to these parts. This is what we refer to as indirect lighting.

The GI system in Unity mimics the rules of how light bounces around in the room. It's the system that calculates how much light gets to the darker parts of the room based on how much it reflects off the surfaces it hits.

Lastly, the lighting mode in Unity is akin to choosing what kind of surfaces you have in the room. Is it a shiny, reflective marble floor or a dark, matte wood floor? Depending on what you choose, the light would bounce off differently and light up the room in a unique way. This is what the lighting mode controls—it determines how the light interacts with different surfaces to create the overall lighting effect in your game scene.

With the GI settings tailored to your project's requirements, it's time to unleash your artistic vision. The last step of the lighting pipeline is to integrate lights, emissive surfaces, reflection probes, and light probes to harmoniously weave your scene's illumination. By following this approach, you will construct a cohesive and detailed lighting environment, enhancing the realism and engagement of your virtual space and ensuring a high-quality user experience.

Now, it is time to implement the steps of the lighting pipeline into our scene. With URP already chosen for your project, let's proceed to the second phase of the lighting pipeline. Here, we'll focus on understanding and selecting the appropriate lighting settings for your scene. This crucial step ensures your virtual world is lit effectively.

Choosing the right lighting settings

As you've selected URP for your XR project, the next task is configuring indirect lighting. This process entails choosing a suitable GI system and lighting mode. **GI** imitates indirect light in 3D environments, leading to soft shadows and a natural appearance as light bounces between objects and surfaces. Unity's lighting mode options present a variety of settings, each balancing visual quality and performance differently.

For simple mobile XR scenes, the most performant option is **Subtractive** mode, which relies on pre-calculated lightmaps to minimize **GPU** load and boost performance. **Shadowmask** mode follows closely, offering **real-time lighting** and baked shadows for static objects, striking a balance between visual quality and performance by reducing GPU load with pre-calculated shadows. Alternatively, **Baked Indirect Lighting** mode provides an ideal mix of real-time lighting, shadowing, and baked indirect lighting for high-quality XR visuals.

To set up indirect lighting, navigate to **Window | Rendering | Lighting** and click on the **New Lighting Settings** button in the **Scene** tab. You'll find the **Lighting** setting within your **Project** panel's `Assets` folder, where it can be renamed. We suggest choosing **Mixed Lighting** by default, enabling the **Baked Global Illumination** option in the **Lighting** settings window. For the lighting mode, **Shadowmask** is the optimal choice for its balance between quality and performance.

> Tip
>
> Though **Baked Global Illumination** permits automatic lightmap generation, we advise against it. Instead, manually generate lightmaps before testing the scene or after implementing changes as it provides the flexibility to test different lighting scenarios and iterate on your design without the overhead of constant automatic updates.

With that settled, you can proceed to place light objects within your scene.

Adding lights and a skybox

Light objects in Unity are an essential part of creating an immersive and believable environment. There are several different light objects to choose from, including Point Lights, Spot Lights, Directional Lights, and Area Lights. They can be categorized under **direct lighting**, which refers to the light that is emitted directly from a light source. Let's take a closer look at each light object.

Point Lights emit light uniformly from a single point, ideal for localized sources such as lamps and candles. **Spot Lights** project light in a conical shape, perfect for directional sources such as flashlights and headlights. **Directional Lights** cast light in a single direction, resembling the sun or moon, while **Area Lights** disperse light from a specific area, akin to ceiling light panels.

Each light type has unique properties, and when combined, they can produce diverse lighting effects. Comprehending these light types is key to achieving effective lighting in Unity.

Let's start by selecting the **Directional Light** GameObject in the **Scene Hierarchy** window and unchecking the checkbox at the top of its **Inspector** window. You'll observe its substantial influence on the scene's illuminance. However, you might be curious why the scene remains brighter than anticipated—the reason lies in the active **skybox**.

In Unity, a skybox serves as a backdrop that surrounds the scene, imparting a sense of environment and depth, similar to the sky and surrounding landscape in real life. A skybox is created using a specific material applied to a large cube that encloses the scene, consistently appearing behind all other elements. This cube generates the illusion of a sky and environment extending beyond the scene's actual boundaries.

Skyboxes play a crucial role in indirect lighting, as they impact the scene's overall lighting ambiance. For instance, a bright sky with a blazing sun casts warm light over the scene, fostering a welcoming and inviting atmosphere. Conversely, a dark sky featuring a full moon casts cool, blue light over the scene, cultivating a more eerie and mysterious mood. Other examples of indirect lighting in Unity encompass GI, ambient light, and lightmaps, which all contribute to the final visual experience.

Let's create and apply a totally dark skybox, following these steps:

1. To create a new material, navigate to the `Materials` folder within the **Project** window. Right-click and choose **Create | Material**.

2. Name the new material `DarkSkybox_Mat` and select it to see its properties in the **Inspector** window.

3. Choose the shader type by going to **Skybox | Procedural** and keep the default settings.

4. Drag the new **Skybox** material to the sky.

Your scene should now resemble the one shown in *Figure 2.11*.

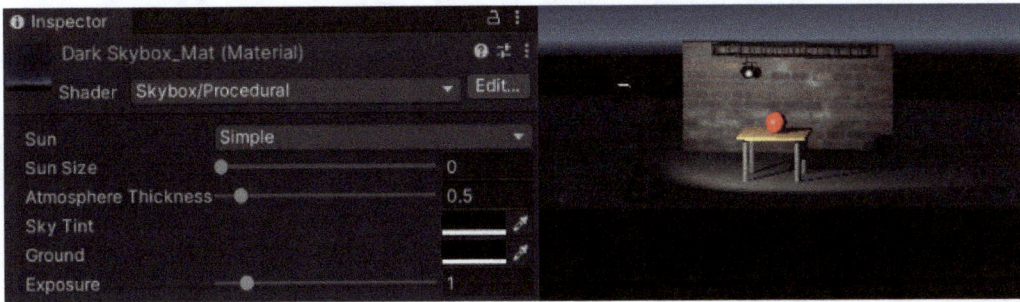

Figure 2.11 – The skybox applied to the scene

Now, you can see that the elements in the scene are only visible because of the activated spotlight emanating from the **ConstructionLightLow** object. Since the influences of indirect lighting conditions have been eliminated, it is time to experiment with the remaining different light types.

First, select the **Spot Light** child object of the **ConstructionLightLow** object and then go to the **Inspector** window. Here, change the light type to **Point**, which, while not entirely suitable for the construction light, will serve to illustrate the shift in behavior. A Point Light, you see, radiates its luminance in all directions, unlike the focused beam of a Spot Light.

Now, let us breathe life into the imported `BrickWallLight` object. To do so, add an empty child object to its existing child object. Rename it `Light` and proceed by selecting **Add Component |** `Light`. With **Spot** as the chosen type, you are free to adjust **Position**, **Color**, **Intensity**, and other parameters to your heart's content. To witness the full extent of your creation's influence, momentarily disable the **ConstructionLightLow** object.

Finally, we have arrived at the last uncharted territory of light types—the Area Light. *Figure 2.12* demonstrates the incorporation of an Area Light into our scene, elevating it to new heights of brilliance.

Figure 2.12 – An Area Light illuminating the scene

To incorporate an Area Light in your scene, go through the following steps:

1. Add a **Cube** object to the scene, scale it to (1.3, 0.75, 0.1), and change its **Position** and **Rotation** parameters so that it is approximately aligned with the white cube light shown in *Figure 2.12*.

2. Name the object Cube Light.

3. Create an empty child object. Right-click and select **Create Empty**, before renaming this fledgling object Area Light.

4. In the **Inspector** window, choose **Add Component** | Light. Now, you will find the option to change the light type to **Area (baked only)**. Temporarily extinguish all other sources of light within your scene by unchecking their respective checkboxes in the **Inspector** window. You may observe that the scene plunges into darkness, with the Area Light seemingly unable to fulfill its illuminating duties.

Unity offers a duo of lighting systems: **real-time lighting** and **precomputed lighting**. Real-time lighting illuminates the scene dynamically, shifting with the player's movements through the environment. In contrast, precomputed lighting calculates the scene's luminance before the game commences, delivering a stable and optimized lighting experience. A third, hybrid mode known as mixed lighting combines the best of both worlds, allowing flexibility in determining which parts of the scene utilize real-time and precomputed lighting.

Area Lights only allow precomputed lighting, requiring us to first bake a lightmap and assign its **emissive surface** to the GameObjects illuminated by the light. Baking a lightmap involves pre-calculating lighting data and storing it in a texture, optimizing the scene's lighting during runtime. Baking is a one-time process that's repeated only when lighting conditions change.

To start, identify the GameObjects affected by the Area Light, such as the Wall object and Cube Light object itself, and activate their **Static** checkboxes in the **Inspector** window. Next, bake the lightmap by navigating to **Window** | **Rendering** | **Lighting**, and switching to the **Baked Lightmaps** tab. Click on **Generate Lighting** to see the transformation.

Notice that only the Wall object appears to shine, while the Cube Light object doesn't. This occurs because there's no object to reflect its light, as with the Ground object below.

To enhance the scene, focus on the Ground, **Sphere**, BrickWallLight, **ConstructionLightLow**, and BrickWallLightStructure objects and their respective children. Enable their **Static** checkboxes and bake the lightmap once more. Observe the transformation of your scene.

Figure 2.13 highlights the difference when using emissive surfaces with Area Lights.

Figure 2.13 – The influence of emissive surfaces of an Area Light

The left image of *Figure 2.13* features only the wall and Area Light as emissive surfaces, while the right image includes multiple GameObjects, resulting in a more dynamic, visually engaging experience.

Let's reactivate the lights of the `BrickWallLight` and the **ConstructionLightLow** objects. By doing so, we embrace the principle of mixed lighting, merging the strengths of both real-time and precomputed lighting. As a result, the objects within our scene become illuminated in a harmonious blend.

As an XR developer, the choice of lighting is a reflection of your artistic vision and the demands of your scene. Real-time lighting captivates with its dynamism, precomputed lighting entrances with its stability, and mixed lighting enchants with its versatility. In the last section of this chapter, we'll introduce light and reflection probes.

Exploring light and reflection probes

Finally, let's delve into **reflection probes** and **light probes** by revisiting our **SampleScene** object. To access it, navigate to **Assets** | **Scenes** in the **Project** window and double-click on the **SampleScene** object.

In Unity, reflection probes and light probes contribute to the creation of realistic lighting and reflections within a scene. Reflection probes capture reflection data from the surrounding environment and apply it to the scene's objects. By capturing a 360-degree view of the environment and storing it as a cube map, reflection probes enable accurate reflections on reflective surfaces in the scene. For instance, if you want a shiny car in a game to reflect nearby trees and the sky, place a reflection probe near the car to gather environmental data and apply it to the car's reflections.

Light probes, conversely, capture lighting data from the surrounding environment and apply it to the scene's objects. By placing probes at various points in the scene, they gather lighting information and store it as a texture. This texture is then utilized to provide realistic lighting to objects in the scene. For example, if a character is standing in a forest, you could place light probes around the character to capture lighting information from the trees and apply it to the character's skin.

Examining the **SampleScene** object in the **Scene Hierarchy** window, we can see it contains three reflection probes and a light probe group with numerous light probes. The images in *Figure 2.14* showcase the appearance of selected reflection probes (left side) and light probes (right side).

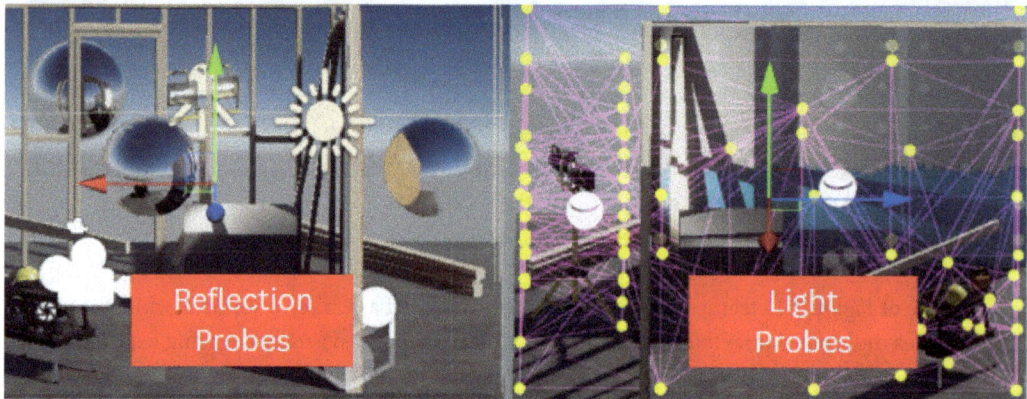

Figure 2.14 – Reflection and light probes in the sample scene

Essentially, not using reflection probes and light probes in a scene may result in less realistic lighting and reflections. However, if your scene doesn't feature highly reflective objects or if the lighting is mostly uniform, removing the probes might not create a noticeable difference in the final result. In the case of this **SampleScene** object, this could very well be true.

As demonstrated, reflection and light probes are potent tools for significantly enhancing a scene's realism in Unity. Nevertheless, their use is not always required or discernible in simpler scenes, as Unity's built-in lightmapping technology is adept at approximating lighting and reflections without resorting to probes.

With a grasp of these principles, you're equipped to create immersive and visually captivating XR experiences.

Summary

In this chapter, you constructed a fundamental scene, familiarized yourself with the Unity Editor, started learning how to import assets to your own project, and were introduced to the essential aspects of lighting.

You embarked on this journey by installing Unity through the Unity Hub and establishing a new project with a sample scene using URP. You then explored the Unity Editor's basics, such as navigating its numerous windows and utilizing the scene editing options, including Grid and Snap. Moreover, you delved into GameObjects and their components, such as colliders.

Subsequently, you crafted your very own rudimentary scene, complete with a ground plane, a brick wall, and a table supporting a sphere. Along the way, you discovered the potency of materials, applied some, and designed your own advanced brick wall material using an imported image as a texture. You also examined the usefulness of prefabs and learned how to transform them into regular GameObjects.

Furthermore, you grasped the importance of importing assets by searching for and importing a spotlight from the Asset Store, addressing potential issues that may arise due to render pipeline incompatibilities. Finally, you delved into the expansive subject of lighting, understanding, and implementing Unity's best practices lighting pipeline within your basic scene. In doing so, you identified the ideal render pipeline, selected the right lighting settings, and experimented with various lights, including Point Lights, Spot Lights, Directional Lights, and Area Lights. You also employed skyboxes and became acquainted with light and reflection probes.

Building upon this solid foundation, you are now prepared to delve deeper into Unity and create increasingly complex and captivating scenes. In the next chapter, we will construct our first VR scene, deploy it, and test it on various VR headsets or simulators, further expanding your understanding and mastery of Unity.

Part 2 –
Interactive XR Applications
with Custom Logic, Animations,
Physics, Sound, and
Visual Effects

After delving into the Unity Engine intricacies and exploring various XR technologies in the first part of this book, this part covers all the essentials of XR development, taking you from a beginner to an intermediate level. In this part, you'll master the creation of your first VR and AR experiences, including deployment to different devices and testing without specific VR headsets or smartphones, using simulators. Once you have a solid understanding of creating and deploying basic XR scenes, you'll progress to developing more advanced XR applications.

You'll add interactivity to your XR applications, utilizing no-code options such as button clicks, and delve into more complex logic by scripting in C#. While this part contains numerous coding segments, don't worry – we'll explain C# in a beginner-friendly manner. Additionally, you'll discover how to enhance the realism of your XR scenes by incorporating sounds and particle effects, while ensuring these physical phenomena adhere to real-world physics laws.

This part contains the following chapters:

- *Chapter 3, VR Development in Unity*
- *Chapter 4, AR Development in Unity*
- *Chapter 5, Building Interactive VR Experiences*
- *Chapter 6, Building Interactive AR Experiences*
- *Chapter 7, Adding Sound and Visual Effects*

3

VR Development in Unity

Let's explore the world of VR, from creating your first VR project in Unity to deploying our first VR scene on a headset or simulator. In this chapter, we'll present the most important VR toolkits and plugins available in Unity, helping you become familiar with each one's capabilities.

You'll gain hands-on experience with the **XR Interaction Toolkit**'s demo scene , understand its most important components and scripts, and learn how to use them in your future VR projects. We'll help you understand the nuances between VR development and traditional game development, as well as share different strategies to ensure a VR headset's computing power.

Another important skill you will gain in this chapter is how to test and deploy VR experiences on various devices, from simulators to VR headsets. This chapter will give you a robust foundation that will equip you with the necessary skills and knowledge to create increasingly complex and immersive VR scenes using Unity.

We'll cover the following topics as we proceed:

- What is VR development?
- Setting up a VR project in Unity and the XR Interaction Toolkit
- Exploring the XR Interaction Toolkit demo scene
- Deploying and testing VR experiences onto different VR platforms or simulators

Technical requirements

To effectively follow along with this chapter on VR development in Unity, it is essential to have an appropriate hardware and software setup. For the most effective and seamless experience, we strongly recommend utilizing a Windows PC or a robust Windows laptop, even if your target platform is a standalone VR headset.

Windows

Windows is the most supported platform by the majority of VR headset manufacturers, including **Meta Quest** and **HTC**. This support stems from Windows' comprehensive hardware support and its optimization for gaming. Whether you're developing for PC-tethered or standalone VR headsets, a Windows environment will provide the most resources and compatibility for VR development in Unity.

For a smooth experience, we recommend the following minimum requirements:

- Operating system: Windows 10 or above

- Processor: Intel i5-4590/AMD Ryzen 5 1500X or greater

- Memory: 8 GB RAM or more

- Graphics: NVIDIA GTX 1060/AMD Radeon RX 480 or greater

macOS

While macOS is rarely supported by VR headsets, Unity still enables VR development on Apple devices through platforms such as ARKit that predominantly target AR instead of VR. If you're using macOS and targeting standalone VR headsets, consider setting up a Windows partition through **Boot Camp**, which allows you to use any Windows-compatible VR headset. This setup will enable you to follow along with the instructions in this chapter without significant issues.

Linux

Linux's VR support is relatively limited, but **Valve** does extend support to the *Valve Index* and *HTC Vive* headsets via SteamVR on Linux platforms. If you're targeting standalone VR headsets such as the *Oculus Quest*, Linux can be used for development by creating Android builds of your VR applications in Unity, then transferring these builds to the headset for testing. However, the standalone VR headsets typically run a version of Android, not Linux, which may bring about unique challenges.

If you're utilizing a Linux platform, please ensure your system meets the following minimum requirements:

- Operating system: Ubuntu 18.04 LTS or newer

- Processor: Dual-core CPU with hyper-threading

- Memory: 8 GB RAM or more

- Graphics: Nvidia GeForce GTX 970, AMD RX480

What is VR development?

When embarking on the fascinating journey of VR development in Unity, it's important that we understand the features and peculiarities of VR development. This includes appreciating the software and hardware limitations, as well as the challenges that exist within this field. In the forthcoming sections, we will dive into the contrasts between VR development and traditional game development for 2D computer screens.

Moreover, we will explore current trajectories and future trends of VR headset technology. This exploration includes the unique pros and cons that each VR headset brings to the table. Such understanding is key as it not only guides you in making a well-informed selection of a VR headset that suits your needs, but also equips you with the necessary knowledge to judge which VR headset would be ideal or less ideal, given a particular context.

By the end of this section, you will be well-equipped to navigate the immersive and dynamic world of VR development in Unity.

Exploring the contrasts between classical and VR game development

Classical game development and VR development, although existing on different ends of the gaming spectrum, both emerged from the same desire to create immersive, virtual landscapes. Intriguingly, both categories share more than just the objective of captivating the audience; they're built on similar foundational elements of game design and require a corresponding toolset.

Both domains of game creation demand careful planning and consideration of gameplay mechanics, user interfaces, objectives, levels, and storylines. Developers, whether they're working on a traditional game or a VR project, often use the same suite of programming languages such as C++, C#, or Python for scriptwriting.

The creation of **game assets** – textures, models, animations, and sounds – is another commonality between the two fields. VR development may lean slightly more toward 3D models and spatial audio to enrich the immersive nature of the environment. However, the overarching process of creating and incorporating these assets remains congruous in both disciplines. Regardless of the gaming medium, be it a 2D screen or a VR headset, the implementation of realistic physics is paramount to an engrossing experience.

Though they share these similarities, the divergences between classical game development and VR development are just as fascinating, primarily due to VR's distinct nature and abilities.

One of the significant contrasts lies in the design's dimensionality. Classical 2D games primarily operate on a flat plane, wherein objects are depicted in two dimensions: height and width. The game *Super Mario Bros.* serves as a classic example, with the action unfolding from left to right and the characters' movements largely restricted within this plane.

Conversely, VR games invite depth as a vital third dimension into play, thus offering the user the freedom to explore the scene from any perspective or position, akin to their experience in the real world. For instance, in a VR game such as *Beat Saber*, players can look around, reach out, and interact with the game environment in a way that mimics real-world spatial interactions.

These subtle yet impactful differences pave the way for a varied range of experiences in classical game development and VR development, each with its own unique charm and challenges.

Equally important, the scope of player movement exhibits a clear distinction between classical 2D games and VR gaming experiences. Traditional 2D games typically restrict player actions to the x and y axes, confining movement within a certain plane. Conversely, VR introduces varying degrees of freedom.

For example, **three degrees of freedom** (**3DoF**) devices, often paired with **Mobile VR** solutions such as *Google Cardboard*, allow the user to look around freely (pitch, yaw, and roll) but don't track physical displacement within a space. This provides a basic level of immersion, enabling the user to view a scene or object from multiple perspectives while standing still.

A step further into the realm of VR are the **six degrees of freedom** (**6DoF**) devices. These not only track a player's view but also register their physical movement along the x, y, and z axes, thus creating a more immersive and interactive environment. Crafting experiences for 6DoF can be intricate, necessitating full 3D spatial interactions. For instance, modern VR headsets such as the *Meta Quest 2* pair up with VR controllers or even offer hand-tracking capabilities. Users can interact with the virtual world in an incredibly intuitive manner, pressing buttons, twisting their hands, and conducting an array of movements that the AI-powered cameras on the VR headset accurately capture and translate in real time.

The imperative of **player safety** is a unique aspect that distinguishes VR development from traditional game creation. While physical safety concerns are virtually non-existent in classical 2D game development, the expanded range of movement in VR can potentially lead to real-world mishaps. As such, developers need to engineer safety features such as virtual boundaries or alert systems to forestall accidents.

The final distinguishing factor lies in **player conditioning**. Classical 2D games generally rely on players' familiarity with standard game controls and mechanics – a safe assumption given the ubiquity of PCs in everyday life. VR, however, often pioneers novel forms of interactions that might confound first-time users. To bridge this gap, developers may need to incorporate tutorials or guides to help players navigate and interact with the VR gaming world.

Not only does VR development diverge significantly from traditional game development, but it also showcases a broader spectrum of techniques to power VR headsets. These diverse approaches, integral to the operation of VR technology, will be unraveled in the next section.

Understanding different approaches to power VR headsets

The power of a virtual reality headset can dramatically impact the success of your VR application. Let's delve into the most common.

PC VR (alternatively known as PC-based VR or tethered VR headsets)

PC-based VR headsets, including models such as *Valve Index* and *HTC Vive Pro*, hinge on the computational capacity of a desktop or laptop computer. They delegate the heavy lifting, such as intensive computations, graphics rendering, and 3D simulations, to the connected computer. To this day, a substantial portion of VR headsets are either wholly PC-based or support a PC VR mode alongside a standalone variant. This is primarily because PC VR, bolstered by the superior computational power and graphical prowess of PCs, provides the most graphically rich and immersive VR experiences. Attributes such as level of detail, frame rate, and responsiveness are top-notch in PC-based VR.

PC-based VR systems predominantly employ two modes of connection between the VR headset and the computer. The wired cable connection, often utilizing HDMI or DisplayPort for video and USB for data, is the traditional route. A more recent innovation is a wireless connection, exemplified by the *Air Link* connection in the *Meta Quest* series. However, wireless PC VR connections are seldom used in reality due to their higher latency, which could hamper the responsiveness of the VR experience. Video quality can also fluctuate depending on network conditions, necessitating a robust and stable Wi-Fi signal for optimal operation.

Despite the convenience of wireless connections, for a seamless VR experience, a wired connection is still advisable, even when the headset supports both. It's essential to note, however, that a wired PC VR system sacrifices portability due to its physical attachment to a PC. Whether you choose wired or wireless, the mobility of your VR setup is limited. For example, you can't easily pack your VR headset for a holiday, an event, a conference, or a friend's gathering. If you decide to transport your VR gear, it necessitates bringing along a powerful PC or ensuring the destination has the required computational resources. Coupled with the potentially steep price of a high-performance or gaming PC, this could make PC VR systems a substantial investment for some users.

Standalone VR headsets

Addressing the limitations of PC-based VR headsets, **standalone VR devices** are making waves in the market, gaining immense popularity. Leading the pack is the *Meta Quest* series. These devices house all necessary computing components within the headsets themselves, making them autonomous units. Most modern standalone headsets are driven by *Qualcomm Snapdragon* chips, offering robust processing capabilities.

Standalone VR headsets champion portability and freedom of movement, eliminating the need to tether to a separate computing device. These all-in-one systems offer a cost-effective alternative to PC-based VR. Moreover, many standalone headsets, such as those in the *Meta Quest* series, support a PC VR mode, providing the flexibility to take on more computation-intensive applications such as *Half-Life Alyx* when desired.

One drawback, however, is the relatively short battery life of standalone headsets, usually topping out around two hours. Nevertheless, remedies exist, such as purchasing comfortable head straps equipped with additional battery life extenders. In most scenarios, we endorse the more affordable standalone headsets such as the *Meta Quest* series over their PC-based counterparts. For academic and business environments, standalone headsets are an efficient choice. A simple solution, such as purchasing multiple standalone headsets or a set of battery-enhancing head straps, can still offer substantial savings over traditional PC VR solutions. Standalone VR headsets also shine when used at events or conferences to showcase products or research, offering a hassle-free and portable solution.

Console-based VR headsets

Console-based VR headsets, such as the *PlayStation VR 2*, cater mainly to the gaming community. They draw on the processing power of the gaming console, delivering high-quality VR experiences without the need for a separate PC. However, the cost of owning a gaming console isn't trivial. Moreover, like their PC-based counterparts, console-based VR headsets sacrifice portability due to their tethered nature. The quality of VR experiences can also be limited by the console's capabilities. In summary, unless your primary focus is gaming, console-based VR headsets may not be the best fit for creating VR experiences, whether in a personal, business, or academic setting. Even within gaming circles, there may be superior options to consider.

Smartphone-based VR headsets

Smartphone-based VR headsets, exemplified by *Google Cardboard* and *Samsung Gear VR*, harness the display and computing power of smartphones, which are devices most people already own. These headsets offer a cost-effective and highly portable solution. However, due to their limited computational and graphical prowess, the immersive experience they provide is inherently diminished. Furthermore, their interaction capabilities within the VR environment are curtailed due to a more restrictive range of sensors compared to other VR headset types. Thus, we seldom recommend them. Even for preliminary testing of VR experiences in the absence of a VR headset, tools such as Unity's XR Device Simulator offer a free, convenient, and even superior solution.

Cloud-based VR headsets

A growing chorus within the VR community is advocating the development of **cloud-based VR headsets**. This emergent technology capitalizes on robust servers to shoulder computational tasks, streaming the results to the VR headset via the internet. This arrangement could potentially deliver high-resolution VR experiences independent of local computational prowess, further enhancing immersion and user satisfaction.

A cloud-based approach could make VR headsets lighter and more user friendly, inviting a broader audience to invest in their own VR hardware. Another appealing aspect is that the hardware housed in the data center that powers the VR headset can be periodically upgraded by the provider, saving users from frequently investing in new equipment.

So, why isn't this computational approach ubiquitous in contemporary VR headsets? The primary reason is the necessity of a stable, high-bandwidth, low-latency internet connection – a prerequisite that isn't universally available. Furthermore, there are potential privacy and security concerns associated with transmitting sensitive data over the internet.

As technology evolves and these challenges are addressed, we may see more adoption of cloud-based VR systems. However, as of now, their use remains largely experimental and not yet mainstream.

Having journeyed through this exploration of the distinct features of VR headsets and the nuances of VR development, you are now well-prepared to embark on the creation of your inaugural VR scene in Unity. This milestone will allow you to play around with some of the engaging toolkits that Unity provides specifically for VR, including the highly versatile XR Interaction Toolkit.

Setting up a VR project in Unity and the XR Interaction Toolkit

In the following sections, you will acquire the knowledge to create a Unity project specifically tailored for VR development. Furthermore, you will become proficient in the installation and configuration of two highly significant plugins that are pivotal to VR development within Unity: **XR Plug-in Management** and the XR Interaction Toolkit.

Creating a project in Unity for VR development

Let's begin by creating a brand-new project in the latest version of Unity. To ensure compatibility with different devices and platforms, it is crucial to enable both **Android** and **Windows/Mac/Linux** build support in Unity. The **Android** build support allows you to test your applications on standalone headsets such as the *Meta Quest* series in standalone mode, whereas the **Windows/Mac/Linux** build support is required for PC-based VR headsets or standalone headsets that are to be used in PC mode. If you haven't downloaded these build supports during the installation of the Unity Editor, you can still do this by opening the Unity Hub and navigating to the `Installs` folder, clicking on the **Settings** icon of your Unity version, selecting **Add Modules**, and installing the missing build supports.

To bring a VR scene to life, we must first navigate to the **Projects** section in Unity and create a fresh project using the latest Unity version. While Unity does provide a VR template among the various project types to choose from, I recommend selecting **3D URP template** instead.

The rationale behind this recommendation is that Unity's VR template does not include the XR Interaction Toolkit, which would lead to extensive setting modifications if you were to choose the VR template. As such, creating a 3D URP project is a more streamlined and trouble-free approach.

In the next section, you will learn how to install XR Plug-in Management and additional XR Plug-ins for PC-based and standalone VR headsets.

Installing XR Plug-in Management and XR Plug-ins

Now, let's head over to **Edit | Project Settings | XR Plug-in Management**. From here, proceed by clicking on **Install XR Plug-in Management**. Think of XR Plug-in Management as the bridge between Unity and the different VR, MR, and AR systems you want to use for your scene. It acts like a middleman, allowing Unity to communicate and work effectively with these devices. Upon successful installation of XR Plug-in Management, you can navigate to **Edit | Project Settings | XR Plug-in Management** to select various plugin providers that will enable your VR application to run in PC VR mode on your VR headset. The following is a brief overview of some of the options:

- **Unity Mock HMD**: Ideal for game developers who may not have direct access to a VR headset. Unity's Mock HMD mimics the functionalities of a VR headset, providing a platform to design, develop, and test your game within Unity's environment without requiring physical hardware.

- **Oculus integration package**: A specialized Unity development package catering to Oculus devices. This includes platform-specific features, Avatar and LipSync SDKs, spatial audio, the Guardian system, and support for hand tracking. However, this package might restrict the portability of your project to non-Oculus platforms.

- **OpenXR**: A remarkable open standard developed by the *Khronos Group* that offers developers the freedom to cater to a broad array of XR devices with the same input, thereby eliminating fragmentation. This obviates the need for separate Oculus or SteamVR plugins, simplifying maintenance by allowing a single code base to operate on any XR system.

If you're an ambitious developer aiming to create a VR experience that appeals to a wide audience, irrespective of their VR hardware, OpenXR would be your go-to. For this reason, we'll be utilizing the OpenXR plugin throughout this book.

To get started, enable the **OpenXR** checkboxes on both the **PC** and **Android** tabs.

This action triggers the download of OpenXR into your Unity project. Once the download is complete, a prompt requesting permission to restart the project will appear. Click **Yes**, as a restart is necessary to transition from Unity's old input mode to the new one. After Unity restarts, you'll find that the **OpenXR** option is now activated in XR Plug-in Management. However, an interaction profile still needs to be added, as indicated by the warning symbol next to it, as shown in *Figure 3.1*.

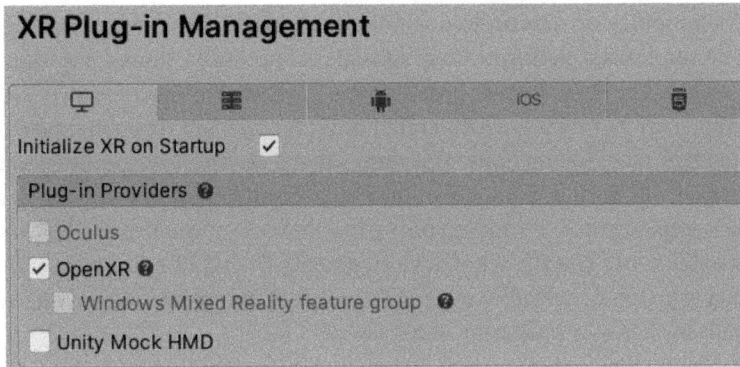

Figure 3.1 – How the OpenXR plugin is selected for the PC tab in XR Plug-in Management

Navigate to **XR Plug-in Management | OpenXR** and click on the + icon in the **Interaction Profiles** section. Here, you can incorporate the interaction profiles of all headsets you aim to support. For instance, if you're using a VR headset from the *Meta Quest* series, you should opt for the **Oculus Touch Controller Profile** option, as shown in *Figure 3.2*.

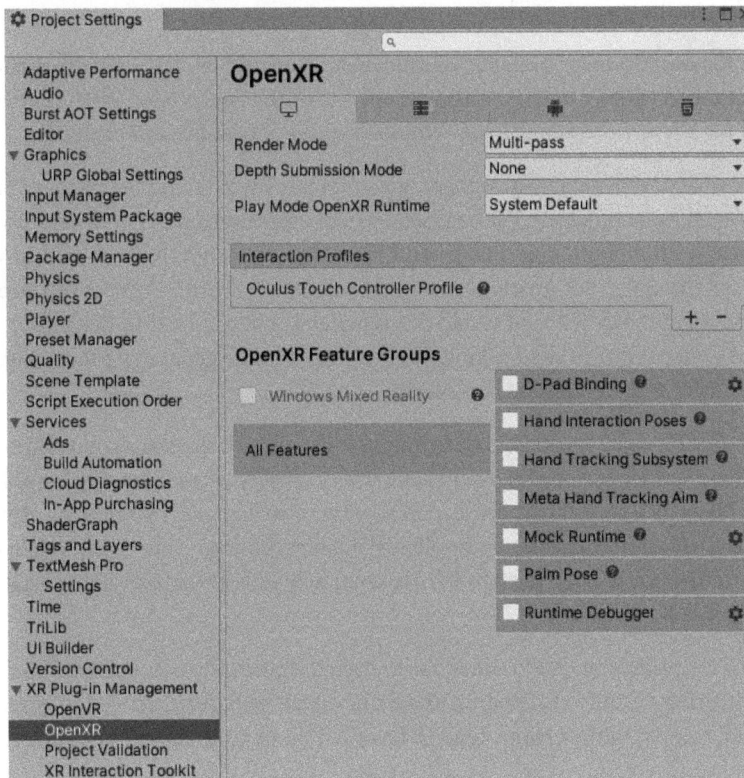

Figure 3.2 – The OpenXR tab with the Oculus Touch Controller Profile selected as an interaction profile

OpenXR's **interaction profiles** are standardized sets of user inputs for XR devices, ensuring consistency and compatibility across various platforms. They allow developers to program once while catering to multiple devices, making the development process more efficient. For instance, two VR controllers from different manufacturers might have different button layouts, but through OpenXR, they can be mapped to a standard set of interactions. This means that if an XR developer programs an action such as *grab* or *jump*, it will work irrespective of the exact controller used as long as the appropriate interaction profile is implemented. By adding only a few interaction profiles, developers limit their application's compatibility to a specific set of devices, potentially excluding some users. Conversely, incorporating as many interaction profiles as possible broadens the application's reach, ensuring that it can be used seamlessly across a wide array of XR devices.

Make sure to add interaction profiles for both **PC** and **Android**. If you can't locate the **Android** tab, this implies that you missed downloading the module during the initial Unity setup. In such a case, you must return to Unity Hub and add the module to your installed Unity version.

You might be wondering why we must configure the same setting twice, once under the **Windows** tab and then under the **Android** tab. The reason is that the **Windows** tab enables VR for PC VR, while the **Android** tab permits the addition of plug-ins from different providers to run your VR application on your VR headset in standalone mode.

Next, let's discuss **Render Mode** for both the **Windows** and **Android** tabs. The default rendering mode is **single pass instanced** (**SPI**), but there's also an option for the **multi pass** (**MP**) rendering mode. Both are tailored for VR environments, each with its own set of benefits and limitations.

SPI offers considerable performance advantages, most notably by reducing the number of draw calls by half. This is crucial for standalone VR headsets, where optimization is paramount to ensure the application runs smoothly. Maintaining a consistent frame rate and smoothness is vital in VR to avoid motion sickness. Additionally, SPI guarantees visual consistency as both eyes are rendered from the same instance, reducing the chances of visual discrepancies that can lead to discomfort. However, the catch is that SPI often requires shader modifications for accurate geometry processing, which can introduce another layer of development complexity.

On the other hand, MP boasts greater shader compatibility since it doesn't demand specific shader modifications. This can be beneficial if your scene uses a variety of shaders. Each eye is rendered independently in MP, which can sometimes lead to superior visual quality by reducing visual artifacts. However, this comes with a performance trade-off: MP doubles the draw calls, which might overwhelm lower-end devices. Furthermore, rendering each eye separately might produce slight visual differences that can be jarring in VR.

As for standalone VR development, SPI is often the preferred choice due to its optimization advantages. It's crucial for delivering visually compelling experiences that also perform well on standalone VR headsets. That said, recent Unity versions tend to favor MP in desktop mode for editor testing.

Having made our project VR-ready, we can now proceed to install the XR Interaction Toolkit and its demo scene.

Installing the XR Interaction Toolkit and samples

By installing XR Plug-in Management, we have enabled Unity to now communicate with our VR headset. But how can we enable Unity to receive information from our headset? And how can our application in Unity translate inputs it gets from the VR headsets into actions in the game? This is where the XR Interaction Toolkit comes into play. To download the XR Interaction Toolkit, do the following:

1. Navigate to **Window | Package Manager**. In **Package Manager**, you can see all the packages you have currently imported, such as the **OpenXR** plugin.

2. Download the XR Interaction Toolkit by clicking the + icon in the top-left corner and selecting **Add package by Name**. Insert com.unity.xr.interaction.toolkit as the **Name** and 2.5.1 as the **Version**.

> **Important note**
>
> Adding a specific version number is optional when installing packages. By default, if you don't specify a version, the Package Manager will automatically install the latest version of the XR Interaction Toolkit. While staying updated with the latest versions is generally advisable for new projects, for the purpose of this book, we strongly recommend sticking to version *2.5.1* of the toolkit. Using this specific version ensures consistency and ease of understanding as you follow along with our tutorials. Even though newer versions might retain a similar structure, the additional features and elements introduced could make navigating our tutorials more challenging. So, to ensure a smooth learning experience, please use version *2.5.1* while working through this book.

---

3. Hit **Enter** and the XR Interaction Toolkit will be automatically downloaded into your project. If you see a pop-up window with a warning asking you to restart the Unity Editor, click on **Yes**. This just means that the XR Interaction Toolkit is using a new input system to validate interactions.

4. Staying in the **XR Interaction Toolkit** tab in **Project Manager**, click the Samples folder to expand a list of useful add-ons to the XR Interaction Toolkit and import **Starter Assets**. This is a collection of tools that make it easier to set up behaviors. It includes a pre-made set of actions you can use as inputs, as well as presets specifically designed for XR Interaction Toolkit behaviors that rely on the input system.

After closing **Project Manager**, you can see that we have a couple more folders in our project now that all relate to the XR Interaction Toolkit. The XR Interaction Toolkit folder is the actual XR Interaction Toolkit, while the Samples folder includes the sample add-ons we downloaded. First, let's open the **Demo Scene** asset that comes with **Starter Assets**. To do this, navigate to **Samples | XR Interaction Toolkit | 2.5.1 | Starter Assets** and double-click on **Demo Scene**.

In the next section, we will explore the different components of the **Demo Scene** asset.

Exploring the XR Interaction Toolkit demo scene

In this section, we will delve into the **Demo Scene** asset of the XR Interaction Toolkit. This scene offers a comprehensive overview of all the crucial concepts you need to grasp to develop your own immersive VR experiences. If you've followed the steps in the previous section, you should now be presented with the scene pictured in *Figure 3.3*.

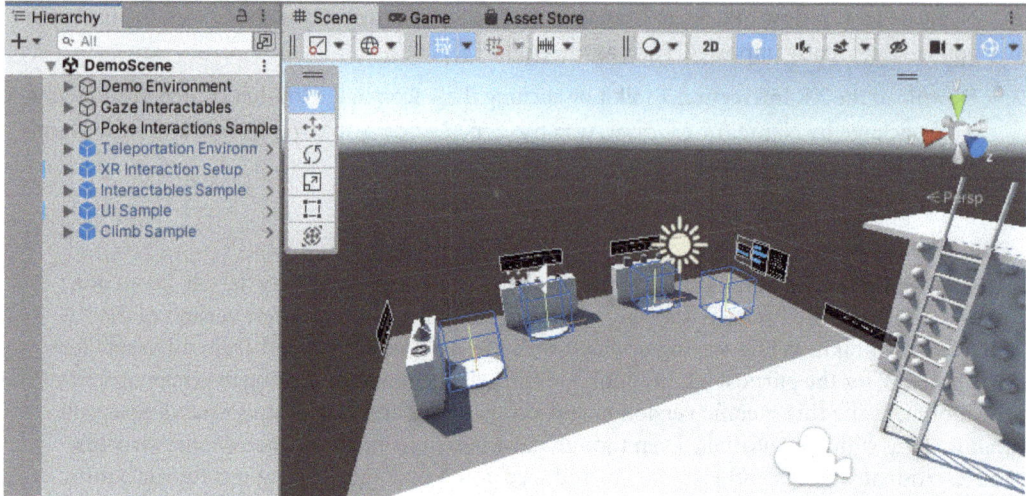

Figure 3.3 – The Scene view and the Scene Hierarchy window of the XR Interaction Toolkit's demo scene

Beyond the **Demo Environment**, **Poke Interactions Sample**, and **Gaze Interactables** GameObjects, you will discover five prefabs (indicated by the blue cubes in *Figure 3.3*) within the demo scene: **Teleportation Environment**, **XR Interaction Setup**, **Interactables Sample**, **UI Sample**, and **Climb Sample**. These prefabs deliver readily usable components and configurations for standard XR behaviors and interactions. For example, if you wish to integrate the **Climb Sample** prefab into another VR project, after importing the XR Interaction Toolkit, simply search for Climb Sample via the project window's search bar. Dragging and dropping this prefab from the project window into the scene hierarchy window will present the identical ladder and climbing wall in your new scene, combined with the associated scripts and functionalities. Similarly, the **XR Interaction Setup** prefab, the most important prefab of the XR Interaction Toolkit, will be a staple of many of the forthcoming XR projects discussed in this book. You will learn more about it in the next section.

Examining the pre-configured player

While the **Teleportation Environment**, **Interactables Sample**, **UI Sample**, and **Climb Sample** prefabs all provide an easy way for you to incorporate different types of interactions into your VR scenes, the **XR Interaction Setup** prefab is the entity that interacts with these interactable components. Essentially, it's a pre-configured player, designed for seamless navigation and interaction within the scene. You can integrate it into any of your future VR scenes in which you import the XR Interaction Toolkit. Let's begin to explore it.

By clicking on the arrow next to the **XR Interaction Setup** prefab in the scene hierarchy window, you can see its child components, which are **Input Action Manager**, **Interaction Manager**, **Event System**, and **XR Origin (XR Rig)**. By opening up all of their child components as well, you can examine the structure of this prefab, as shown in *Figure 3.4*.

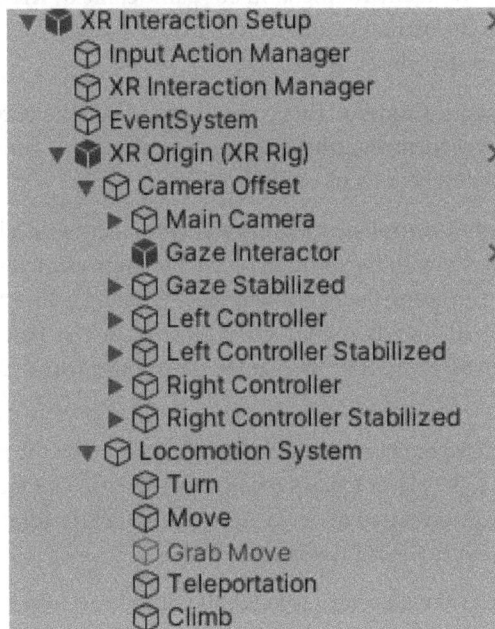

Figure 3.4 – The components of the complete XR Interaction Setup prefab

To understand these components and how they interact with each other, let's imagine a VR game scenario where the player takes on the role of a chef in a restaurant and uses a VR headset and controllers to interact with the kitchen environment. In this imaginary example, the components of the **XR Interaction Setup** prefab would take on the following roles:

- **Input Action Manager**: This component manages the processing of inputs from the VR controllers. For instance, a controller button press could initiate actions such as gripping a pan or picking up an ingredient.

- **XR Interaction Manager**: This component governs interactions between the player and virtual objects. For instance, it manages the outcomes when a player's VR hand interacts with a virtual button – perhaps a door opens or a light switches on. Similarly, if the player, holding a pan, moves their hand over the stove and releases the grip button, the pan could settle onto the stove.

- **EventSystem**: This component oversees the triggering and processing of events within the game. For example, upon the successful preparation and serving of a dish, an event could be triggered perhaps the sound of applause or a visual congratulatory message illuminating the screen.

- **XR Origin (XR Rig)**: This component serves as the player's anchor point in the virtual world, responsible for managing the user's movement and positioning within the XR environment. By utilizing data from the VR device, **XR Origin** updates the position and orientation of the camera and controllers via its attached **XR Origin** script, thus creating the illusion of movement and interaction within the virtual environment. In our game context, **XR Origin** would represent the player's location in the virtual kitchen; if the player advances a step, their avatar would correspondingly move a step closer to the stove.

- **Camera Offset** with **Main Camera**: This component tracks the player's head movements. For instance, if the player turns their head toward a recipe book lying on a shelf, the game's viewpoint adjusts accordingly to focus on the book.

- **Left Controller** and **Right Controller**: These components monitor the player's hand movements, which typically correspond with the position of the VR controllers. If the player extends their hand to grab a virtual pan using the controller, the VR hands within the scene would mimic this action. This is why the hands contain child objects such as **Poke Interactor** for touch, **Direct Interactor** for grabbing, **Ray Interactor** for pointing, and **Teleport Interactor** for swift movement.

- **Gaze Interactor**: This component facilitates interactions based on the player's eye or head gaze direction. For instance, if the player focuses their gaze on a particular ingredient, a tooltip might appear to deliver more information about it. As not every VR headset supports eye tracking, **Gaze Interactor** also supports head tracking.

- **Locomotion System**: As its name already suggests, this GameObject with its associated **Locomotion System** script is responsible for equipping the player with different forms of movement such as turning, walking, teleporting, or climbing. Let's have a look at each of its children:

  - **Turn**: The **Turn** GameObject has two important scripts attached to it, which you can see by selecting **Turn** in the scene hierarchy window and navigating to its inspector window. First, there is the **Snap Turn Provider** script. This script enables the player to rotate their view in the VR environment in discrete steps, or "snaps," using their VR controllers. This can be useful in cases where physically turning isn't convenient or possible.

  - Second, there is the **Continuous Turn Provider** script. Allowing for smooth, continuous turning in the VR environment using the VR controllers, this script offers an alternative to the step-by-step turning provided by the **Snap Turn Provider** script.

- **Move**: In our VR restaurant game, for example, the player can move through the kitchen by navigating with the controllers of their VR headset. Even though they remain stationary in the physical world, their perspective shifts in the virtual space, giving the illusion of traversing the environment as if they were genuinely walking through the virtual kitchen. This functionality comes from the **Dynamic Move Provider** script, which is attached to the **Move** GameObject.

- **Grab Move**: This component of the XR Interaction Toolkit's **XR Interaction Setup** prefab facilitates intuitive movement within the VR space. With two instances of the **Grab Move Provider** script being attached to this GameObject, the player can simulate the sensation of grabbing the virtual world with either hand. As the player moves the controller, the VR origin adjusts inversely, keeping the controller's position consistent relative to the virtual environment. Complementing this, the **Two-Handed Grab Move Provider** script empowers players to manipulate their position using both controllers, similar to grabbing two bars and pulling or pushing oneself. This dual-hand interaction allows players to not just move through the VR restaurant game, but also to adjust their orientation and scale within the scene.

- **Teleportation**: Solely navigating the virtual world via continuous movement, as provided by the **Move** GameObject, can make some VR users motion sick, even if they regularly immerse themselves in VR. This is why most VR applications mainly work with teleportation, which enables the player to instantly appear in a new spot within the virtual space by moving the joysticks of the VR controllers in the desired direction. As they do so, a visual cue, often a circled overlay, indicates the target destination. Once the joystick is released, the player is directly teleported to that location. The **Teleportation** GameObject of the XR Interaction Toolkit offers the player of our imaginary VR restaurant game a rich variety of ways to teleport around the kitchen or dining area. For example, the player might only be able to navigate to distinct places in the VR restaurant, making the gameplay more straightforward and structured, which is ideal for educational settings or applications tailored to newbies to VR. If exploration and spontaneity are the core missions of the VR restaurant game, enabling the player to teleport to any part of the floor that is not occupied by chairs, tables, or kitchen cabinets might be more favorable.

Given the importance of teleportation in VR applications, we'll delve deeper into which teleportation features the XR Interaction Toolkit offers in the upcoming section.

Teleporting

The **Teleportation Environment** prefab of the **Demo Scene** asset consists of a **Teleport Area** prefab and four **Teleportation Anchor** prefabs. You can see these components in *Figure 3.5*.

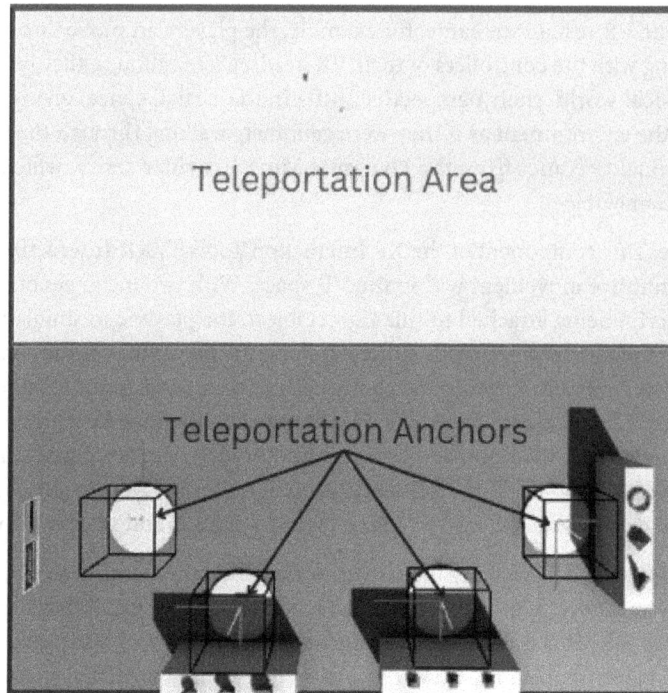

Figure 3.5 – The Teleportation Area and four Teleportation Anchors, which are both part of the Teleportation Environment prefab

The **Teleport Area** script, equipped with a **Teleportation Area** prefab and a child cube, represents a space available for teleportation. Users can teleport to any pointed spot within this designated area, effectively teleporting within the bounds of the cube.

By contrast, a **Teleport Anchor** prefab serves as a specific teleportation point, allowing users to navigate to exact locations with ease.

Imagine a VR game where users must traverse a river by hopping from one stone lying in the water to another. On the other side of the river, there is a forest with gold hidden underneath some trees that they must find. In this case, **Teleport Anchors** could be used for each of the stones lying in the water. This would allow the users to traverse the river with ease and focus more on the actual gameplay. For the other side of the river, however, a **Teleport Area** script would be better suited, allowing users to explore the area as they wish.

In the next section, you will learn all about grabbable objects and their movement types.

Exploring grabbable objects

The **Interactables Sample** prefab, another part of the **Demo Scene** asset, contains three 3D figures that highlight the capabilities of the **XR Grab Interactable** script, an essential tool within the XR Interaction Toolkit. This script will likely be a staple of your future VR projects whenever you want an object to be grabbable. Once you've integrated the XR Interaction Toolkit and configured XR Plug-in Management in any of your VR endeavors, making an object grabbable is a breeze. Simply highlight the object in the scene hierarchy window, click on the **Add Component** option in the Inspector, and then search for and choose the **XR Grab Interactable** script. Additionally, ensure the **XR Interaction Setup** prefab is in your scene, as it provides the required player for the interaction.

This single step of adding the **XR Grab Interactable** script to an object transforms it into a VR-interactable item. This ease of use and efficiency is a significant reason behind the XR Interaction Toolkit's popularity among VR developers. Instead of manually coding numerous lines to achieve this functionality, you can focus your time on refining the user experience by tweaking the integrated physics settings. These adjustments can significantly enhance the immersion and realism of your VR application.

Now, let's delve into the various physics configurations offered by the **XR Grab Interactable** script to enhance the object-grabbing experience. Each of the three figures of the **Interactables Sample** prefab showcases a different **Movement Type** setting of the **XR Grab Interactable** script, as you can see in *Figure 3.6*.

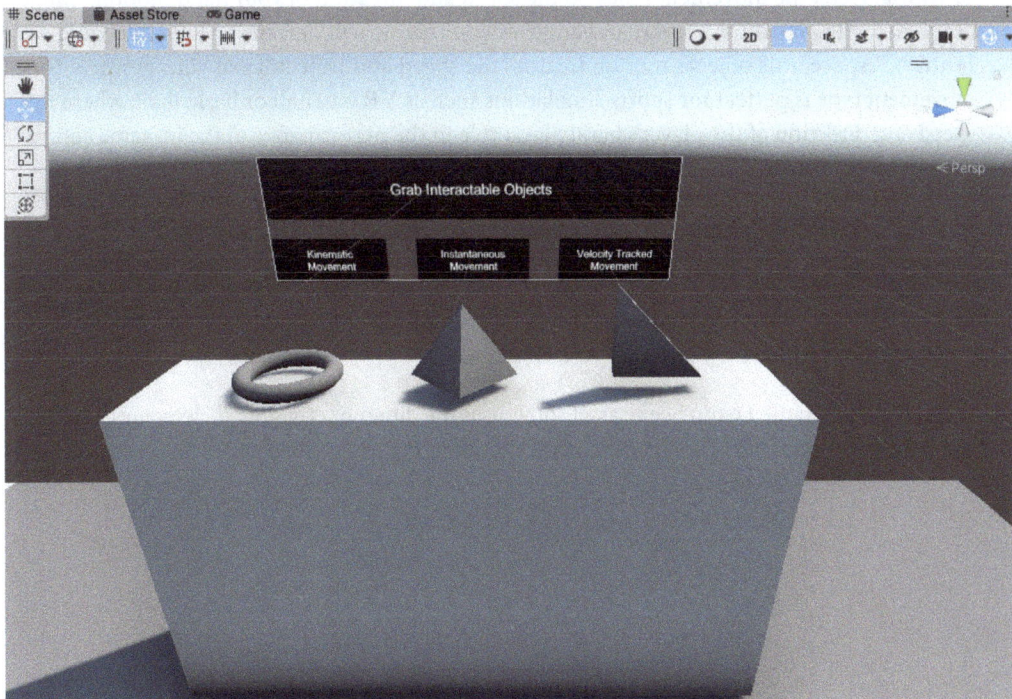

Figure 3.6 – The demo scene's Interactables Sample prefab in the scene view

Let's go through the figures one by one:

- **Interactable Kinematic Torus**: When you select this in the scene hierarchy window and inspect its properties, you will see that its **Movement Type** setting is labeled as **Kinematic**. A kinematic object is guided by the software's calculations rather than by external forces. This means it moves smoothly based on the game engine's instructions and isn't affected by gravity or collisions like a regular object would be. This type of movement can be useful in applications such as puzzle games, where objects must fit into specific locations without bouncing or rolling away. It ensures that once placed, the objects remain static.

- **Interactable Instant Pyramid**: This figure has **Instantaneous** selected as its **Movement Type** setting. Instantaneous movement is precisely as it sounds: immediate. Instead of creating a sensation of weight or inertia as you move it, this object will respond and relocate right away without any noticeable lag or drift. It's like moving a cursor on a computer screen; where you point, it goes. This is ideal for **user interface** (**UI**) interactions in VR, such as selecting menu options or dragging and dropping virtual files. It's also useful in time-sensitive scenarios, such as a fast-paced game where players must rapidly move objects to different locations without the delay of simulated physical interaction.

- **Interactable Velocity Tracked Wedge**: This figure's **Movement Type** setting is set to **Velocity Tracking**. Here, the movement relies on the speed and direction of your hand motion. Think of it like tossing a ball; the speed and angle of your throw will determine how the ball flies. In the VR space, this object's motion mirrors the speed and trajectory of your gesture. This movement type is perfect for sports simulations such as VR baseball or basketball, where the speed and direction of the player's hand play a role in the performance of the in-game action.

You can observe these three movement types firsthand by running the demo scene on your VR headset or via the XR Device Simulator and navigating to the **Grab Interactable Objects** booth.

The next section introduces you to the different types of UI elements provided by the XR Interaction Toolkit.

Inspecting different types of UI elements

The **UI Sample** prefab in the **Demo Scene** asset showcases the different types of UI elements currently provided by the XR Interaction Toolkit. You can see them in *Figure 3.7*.

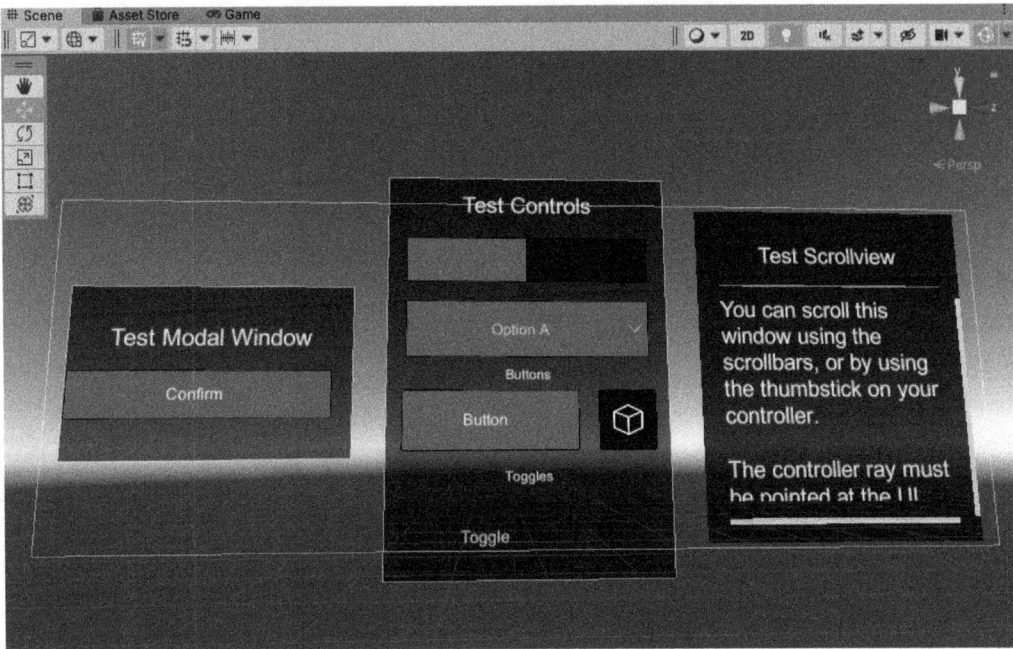

Figure 3.7 – The demo scene's UI Sample prefab in the scene view

Click on the arrow next to the prefab in the scene hierarchy window to inspect its child components. Let's go through them:

- **ModalSingleButton**: This UI element presents a short text message accompanied by a button. This layout makes it an ideal UI prefab whenever you need to display important pop-up notifications to the user of your VR experience. For instance, it could serve to inform users at the outset of a VR session that they're about to embark on a timed experience. Alternatively, it can ensure that users have understood instructions or confirm their consent for data usage in a VR research study.

- **Interactive Controls**: This prefab contains multiple UI elements with more detailed input choices for users compared to the **ModalSingleButton** prefab. **MinMaxSlider**, for instance, lets the user pinpoint a specific state within a minimum-to-maximum range by sliding over it with the VR controller. The **Dropdown** prefab, **Text Toggle** prefab, and **Icon Toggle** prefab grant similarly precise ways for users to communicate their preferences in the VR world. These components of the **Interactive Controls** prefab shine in contexts requiring intricate user feedback or directives. For instance, the user might select the room they want to see in a VR apartment viewing experience via a **Dropdown** menu. They could adjust the room lighting intensity with **MinMaxSlider**, and toggle between floor materials and wall paintings using **Text Toggle** and **Icon Toggle**.

- **Scroll UI Sample**: This prefab features a simple text block that users can scroll through using their VR controllers. This element is a handy tool for educational segments within VR, such as diving deep into the history of an art piece in a virtual museum tour.

By right-clicking on the UI sample prefab in the scene hierarchy window and selecting one of the options offered under UI, you can easily add more UI elements onto the **Canvas** component as you please.

The **TrackedDeviceGraphicRaycaster** script is another essential component of all UI interactions in VR scenes with the XR Interaction Toolkit. It should be attached to the **Canvas** component showcasing the UI elements. In the case of the **Demo Scene** asset, the **UI Sample** prefab is the **Canvas** component, so the script is attached to it, as you can also see in its inspector window. The **TrackedDeviceGraphicRaycaster** script gives your **Canvas** component an X-ray vision to discern when a player's controller is aiming at an element on the **Canvas** component. Within the VR context, it transforms controller movements into interactions with the UI elements displayed on the **Canvas** component.

In the next section, you will learn how you can interact with buttons in your VR scene.

Interacting via buttons

The **Poke Interactions Sample** GameObject within our **Demo Scene** asset features three unique buttons that are shown in *Figure 3.8*.

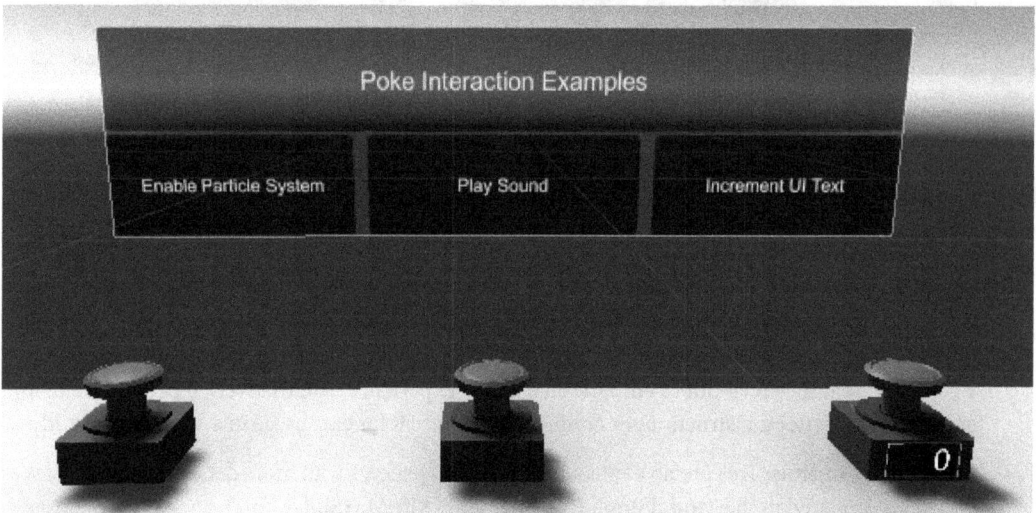

Figure 3.8 – Three unique buttons from the Poke Interaction Sample of the demo scene

Unlike ordinary buttons, these are designed to react to your touch in distinct ways. The button on the left side triggers a particle animation once it is poked; the button in the middle plays a sound, and the button on the right side increments the number on the UI element that is attached to it every time it is pressed.

These buttons all have the same three scripts attached to them: **XR Simple Interactable**, **XR Poke Filter**, and **XR Poke Interactor**. These scripts provide important functionalities that all of these buttons need, despite their distinct flavors:

- **XR Simple Interactable**: This script is the magic that makes our buttons interactive and touch-responsive. Without it, they would simply be static objects in the virtual world, unreactive to our touch. In the scene hierarchy window, select the first **Push** button and navigate to the inspector window. Now, click on the arrow next to the **Interactable Events** menu of the **XR Simple Interactable** script to see all interactable events associated with this script. Scroll down to the **Select** section and observe how the **Particle System** event is played or stopped as the user enters or exits the selection of this button. Repeat this step for the other two buttons to see what their **Select** sections look like. In the upcoming chapters of this book, you will create your own unique game logics in the **Interactable Events** section.

- **XR Poke Filter**: This script assesses the authenticity of a touch. If the touch doesn't meet its criteria, the poke filter blocks its entry.

- **XR Poke Follow Affordance**: This script provides the buttons with a lifelike, tactile response. Upon poking, the respective button is pressed down, mirroring the movement of a real-world button. Upon release, it rebounds back to its original position.

Throughout this book, we'll delve deeper into creating diverse poke interactions, teaching you to customize these components to fulfill your specific requirements.

On our last step of exploring the **Demo Scene** asset, you'll get a glimpse of more advanced features of the XR Interaction Toolkit.

Exploring gaze interactions and climbing

Gaze interaction is an immersive feature in VR that enables the user to trigger events and control aspects of the VR environment simply by directing their eye or head gaze. The demo scene of the XR Interaction Toolkit contains an entire booth just to showcase different types of gaze-based interactions. These types of interactions are among the most advanced features of the XR Interaction Toolkit, which is why you will learn how to implement them into your VR scenes in *Chapter 8*, alongside other advanced techniques.

Climbing, while a demanding concept in real life, is easier to enable in VR scenes than one might think, especially when compared to techniques such as gaze tracking. The **Demo Scene** asset's **Climb Sample** prefab showcases two climbable objects: a ladder and a climbing wall. You can observe them in *Figure 3.9*.

Figure 3.9 – The Climb Sample prefab, consisting of a ladder and a climbing wall among other elements

At first glance, they may seem distinct, but underneath, they share the same scripts.

To understand how you can make an object in your VR scene climbable, follow these steps:

1. Open the **Climb Sample** prefab in the scene hierarchy window.

2. Explore every child component of both the **Ladder** and the **Climbing Wall** prefab that has the *Climbable* label. In the inspector window, you'll notice each of these climbable components has the same two scripts from the XR Interaction Toolkit attached to them: the **Climb Interactable** script and the **XR Interactable Affordance State Provider** script.

Simply adding these two scripts to any object in your VR environment via the inspector window provides it with climbable properties. Now, let's dive deeper into the roles of these two scripts:

- **Climb Interactable**: This script essentially makes an object climbable. When a player interacts with an object that has this script, they can virtually climb it.

- **XR Interactable Affordance State Provider**: This script aids in fine-tuning the interaction between the player and climbable objects. It's a system that dictates how the climbing experience should function and respond to the user's interactions.

Now that we've explored all the key components of the demo scene, it's time for you to test and explore it. In the upcoming section, we'll discuss several options to get you started on this exciting journey.

Deploying and testing VR experiences onto different VR platforms or simulators

Creating a VR scene goes beyond just design; it involves testing and deployment as well. While testing focuses on ensuring the VR scene works correctly and offers a good user experience, deployment is about transferring and optimizing the scene for playback on specific VR headsets, such as the *Meta Quest* series, *HTC Vive*, or *Valve Index*. Early development stages might use tools such as the XR Device Simulator for rapid testing, but as a project nears completion, thorough testing on an actual VR headset becomes essential.

Now, let's go through the various options for testing and deploying your VR scene based on what hardware is available to you:

- *No VR headset available*: In cases where a VR headset is unavailable, you can use the XR Interaction Toolkit's XR Device Simulator to test the features of your VR scene. Even if you do have a VR headset at your disposal, the simulator can offer a quicker alternative to physically connecting and testing on a VR headset, particularly when you're making simple modifications.

- *Standalone VR headset*: Most standalone VR headsets support PC-based VR mode. This mode, when enabled, is often the most efficient and powerful way to test your VR scene. However, there might be times when you don't have the necessary cable to activate PC-based VR mode. In such cases, you have two main alternatives for testing your VR scene. The XR Device Simulator is typically the preferred method, because testing directly on a standalone VR headset requires deploying it onto the device. This process can be lengthy and time-consuming. It happens on the Android platform via the project's *Build Settings*. So, if PC-based VR mode isn't an option, you might use the XR Device Simulator for quick tweaks and minor scene changes. For substantial modifications or a comprehensive user experience test, deploying directly to the headset would be advisable.

- *PC-based VR headset or Standalone VR headset with PC-based VR mode enabled*: If you have either a PC-based VR headset or a standalone VR headset with PC-based VR mode enabled, your options for testing and deployment are expanded. For minor modifications to your VR scene, the XR Device Simulator might be all you need. For moderate to major changes, it is easy to test the VR scene on your headset by pressing the **Play** button in your Unity scene, provided the VR headset is connected to either the *SteamVR* or *Oculus* software. For considerable changes, or when analyzing the overall user experience, you can deploy your VR scene through **Windows/Mac/Linux** in **Build Settings** while the VR headset is connected to the *SteamVR* or *Oculus* software.

You might be curious about what *SteamVR* or *Oculus* software entails and when you'll need them. Simply put, both are software platforms that not only let you play PC VR games but also test and deploy VR experiences on your headset. You will learn more about both in the upcoming sections of this chapter. In general, we would recommend the following, depending on the type of headset you own:

- *Headset from the Meta Quest series*: Install both SteamVR and Oculus software on your computer and learn to deploy your VR scene to your VR headset through both VR platforms, as these headsets support both. Both types of deployment are very easy to do, but they enable you not only to test your scene more thoroughly, but also to easily reach a wider audience.

- *Valve Index, HTC Vive series, or any other non-Meta headset*: Only install SteamVR and learn to deploy your VR scene onto your VR headset.

The following sections will delve deeper into each of these options, providing a detailed guide on installing the XR Device Simulator, SteamVR, and Oculus software. Feel free to skip to the sections that are relevant to you.

Installing the XR Device Simulator

Now, we'll learn about the essentials of installing and using the XR Device Simulator, a valuable tool in the VR development workflow that lets you emulate user inputs to control XR devices in a simulated environment. From importing the toolkit to using it effectively in your Unity projects, we'll detail every aspect using comprehensive examples.

As it is part of the XR Interaction Toolkit, you can easily import the XR Device Simulator via Unity's package manager. Here is a step-by-step guide to doing this:

1. Inside your Unity project, navigate to **Windows | Package Manager | XR Interaction Toolkit**.
2. By selecting the **Samples** tab of the **XR Interaction Toolkit** package, you will find **XR Device Simulator** among the listed elements.
3. By clicking the **Import** button next to the XR Device Simulator, you import it into your project.

After successfully installing the XR Device Simulator, let's use it to test out the **Demo Scene** asset of the XR Interaction Toolkit.

Using the XR Device Simulator

To incorporate the XR Device Simulator into the demo scene or any of your future VR scenes, search for XR Device Simulator via the project window's search bar. This way, you will find the **XR Device Simulator** prefab, the asset you will need to add to scenes where you want to simulate XR input. Now, simply drag this prefab into the scene hierarchy window of your VR scene. Before continuing, make sure the **XR Interaction Setup** prefab is also added to your scene. Now, click on the **Play** button to explore the power of the XR Device Simulator. *Figure 3.10* shows what the XR Device Simulator looks like in the **Play** mode.

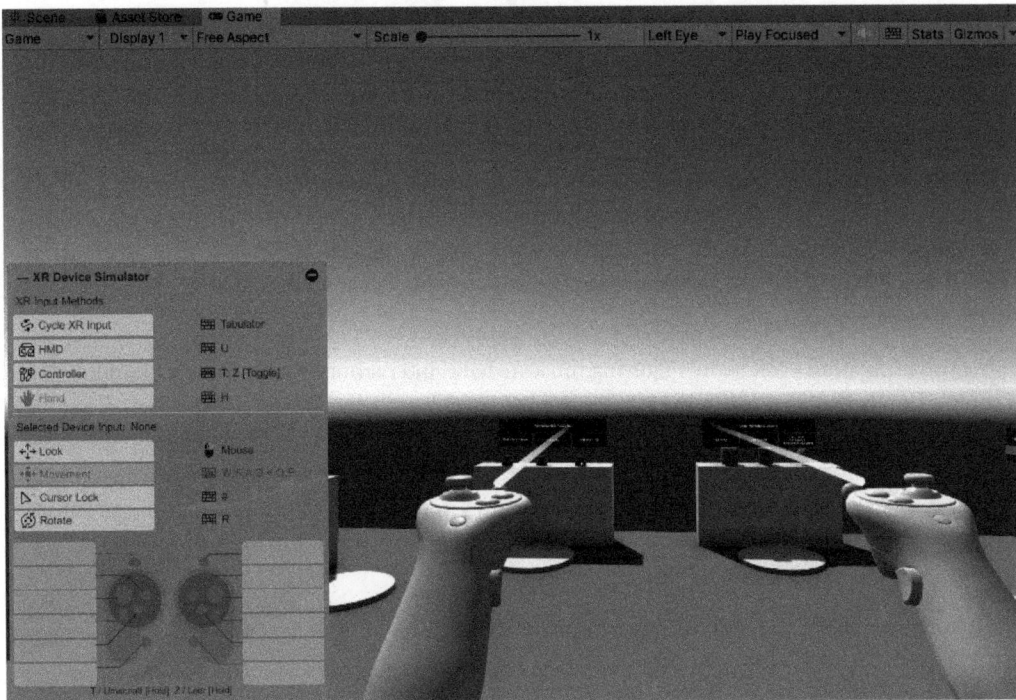

Figure 3.10 – The Play mode of the XR Interaction's Toolkit demo scene
when the XR Device Simulator is added to the scene

You are now equipped to navigate your scene using the key bindings supplied by the simulator, specifically the *WASD* keys. Hit the *Tab* key to switch active control between the left controller, right controller, and VR headset. Utilizing your mouse, you can explore the scene or alter the angles of your controllers, contingent on whether you are operating in headset or controller mode. Pressing the *G* key enables you to interact with objects in the scene, including pressing buttons. After experimenting with the XR Device Simulator and familiarizing yourself with the different keys, you'll swiftly appreciate its simplicity and effectiveness as a testing tool for your VR scene before moving on to real XR hardware deployment.

As powerful and convenient as the XR Device Simulator might be, there is no experience more immersive than testing and deploying your scene onto a VR headset. The following section explains how to do so using SteamVR.

Setting up SteamVR

SteamVR is a platform developed by *Valve Corporation* for VR applications and content. It is part of the larger **Steam** platform, a very popular online marketplace for games and software, known for its extensive library and active community. For VR developers, using SteamVR can thus be very advantageous as

VR content in SteamVR can potentially reach millions of users worldwide. SteamVR supports a wide range of VR headsets, including those from *HTC Vive, Valve Index, Windows Mixed Reality*, and the *Meta Quest* series. At the time of writing the book, we are not aware of any commercially available VR headset that doesn't support SteamVR. It is basically a default standard in the VR headset market.

Interestingly, even the *Meta Quest* series, designed on Android and not primarily PC-based VR, allows developers to deploy VR scenes and applications not just on Meta's proprietary Oculus platform, but also on the SteamVR platform. This versatility is immensely beneficial. If you're a *Meta Quest* series headset owner, we highly recommend learning to test and deploy your scenes on both platforms. Let's illustrate this with an example.

Suppose you're creating a VR game requiring intricate hand interactions such as object manipulation and complex maneuvers. You aim to cater to users of *HTC Vive, Valve Index*, and headsets from the *Meta Quest* series. By harnessing SteamVR and its SteamVR Unity plugin, you can develop the game once and ensure compatibility across all these headsets. SteamVR's precise tracking system enhances the game's immersive and responsive nature across various devices. Post-development, you can launch your game on the Steam platform, thus accessing a vast user base, including users of the *Meta Quest* series who are employing SteamVR.

In the following section, you will learn how to install SteamVR.

Installing SteamVR

To target SteamVR, you must first install and setup SteamVR. To do this, perform the following steps:

1. Download Steam to your PC from the official website (`https://store.steampowered.com/about/`), create a Steam account, and log into this Steam account via the Steam software.

2. Inside Steam, search for `SteamVR` in the Steam Store, select the **Play Game** button, and follow through with the on-screen instructions to begin installing SteamVR.

3. Once the installation of SteamVR is completed, connect your headset to your PC via a cable or a wireless connection such as *Air Link* for the *Meta Quest* series. Then, either search for the SteamVR plugin inside the Steam software and click the **Launch** button, or directly open the SteamVR software that was downloaded onto your PC by searching for `SteamVR` in your regular PC apps and clicking on it. Either way, you should quickly see that your VR headset is connected to SteamVR on your PC. Inside your VR headset, you should now see the SteamVR environment and game store.

While some VR headsets, such as *Valve Index*, only necessitate the installation of SteamVR software for testing and deploying VR experiences, many other headset manufacturers demand additional software to be installed alongside SteamVR. For instance, *Windows Mixed Reality* devices, including the *HP Reverb* series, also require the *Mixed Reality Portal*. These installations are typically straightforward and usually don't involve any further steps from you other than opening the downloaded software every time you want to connect your VR headset to the PC. As these software requirements can change over time, we recommend you check whether any manufacturer-specific software is needed for your particular VR headset model to test or deploy on it.

After installing and setting up SteamVR and potentially some other manufacturer-specific software, the next section reveals how you can finally test and deploy your scene onto your VR headset.

Testing and deploying your VR scene with SteamVR

When developing for SteamVR in Unity, you can test your VR scene directly in the Editor using the **Play** button. With the SteamVR plugin installed, the Unity Editor will recognize a connected SteamVR-compatible headset, and the scene will be rendered in VR. This enables you to quickly evaluate changes, troubleshoot issues, and refine your VR content without leaving the Unity environment.

When you're ready to deploy your scene, you can build the project into an executable file that can be run outside of Unity.

After connecting your VR headset to SteamVR, go back into your Unity project. Here, go to **File | Build Settings** in your Unity project. Make sure that your preferred **Target Platform** field, such as Windows, is selected and that the **Architecture** field is set to 64-bit. Finally, you can deploy your VR scene onto your headset by clicking the **Build and Run** button in **Build Settings**. Once built, the project can be launched in SteamVR through the Steam client, allowing you to experience the VR scene on your headset as end users would.

If you own a VR headset from the *Meta Quest* series, setting up your VR scene for Oculus will be just as important to you as setting it up for SteamVR. The next section explains how to do this.

Setting up Oculus

In this section, we'll explore the steps for setting up Oculus for your VR development projects. We'll detail the installation of the Oculus app on your PC, how to test your VR scenes with Oculus, and how to deploy your VR scenes with Oculus.

Installing Oculus

To work with Oculus in Unity, you'll first need to download and install the Oculus app onto your PC. Complete the following steps:

1. Navigate to www.oculus.com/setup in your web browser.
2. Below the VR headset of your choice, such as *Meta Quest Pro*, click **Download Software**.
3. Open the downloaded Oculus app and click the **Install Now** button.
4. Follow the on-screen instructions to create an Oculus account and set up your VR headset.

After the installation and setup are completed, it is finally time to test and deploy your VR scene with Oculus. This process is described in the next section.

Testing and deploying your VR scene with Oculus

Once the Oculus software is installed on your PC, you can test your VR scenes directly in the Unity Editor. The following steps show you how:

1. Connect your Oculus-compatible VR headset to your PC. Ensure the Oculus app recognizes it.
2. Open your VR project in Unity.
3. Click the **Play** button in the Unity Editor to start the scene. Unity will automatically render the scene in VR, allowing you to interact with the scene using your Oculus headset.

After testing and refining your VR scene, you can build and deploy it so it can be experienced outside of Unity. Complete the following steps:

1. With your VR project open in Unity, go to **File | Build Settings**.
2. Make sure your preferred **Target Platform** field, such as Windows, is selected.
3. Ensure the **Architecture** is set to 64-bit.

Click the **Build and Run** button. Unity will build an executable file of your project and run it. After the build is complete, the executable can be launched with the Oculus software. Now, you can fully experience your VR scene with your Oculus headset, just as your end users will.

Summary

Throughout this chapter, you've navigated the exciting landscape of VR development in Unity. You've learned to create a fundamental VR scene from scratch and mastered the deployment process on various devices. The utilization of the XR Interaction Toolkit and the informative elements from its demo scene should now be within your repertoire, empowering you to generate basic VR scenes with increased confidence and proficiency.

Furthermore, these acquired skills aren't limited to a specific genre. Whether it be for industrial applications or academic research, your new proficiency in VR development should allow you to replicate this process effectively to fit any use case that you encounter.

As we venture into the next chapter, we will broaden our horizon to include the creation and deployment of AR scenes within Unity.

4

AR Development in Unity

In this chapter, we will immerse ourselves in the fascinating realm of AR development, from creating our first AR project in Unity to launching our first AR scene on a device or simulator. We will present to you numerous AR toolkits and plugins that Unity offers, and guide you in understanding their unique functionalities.

In a step-by-step manner, we will walk through the process of establishing an AR project in Unity, ensuring it is primed for smooth deployment onto any AR-supportive device.

This chapter will cover the following topics:

- Understanding the AR landscape
- Setting up an AR project in Unity using AR Foundation
- Testing AR experiences directly in Unity
- Deploying AR experiences onto mobile devices

Technical requirements

Before we dive into the practicality of the Unity Editor, it's important to ensure your computer system is up to the task. *Unity 2021.3 LTS*, or a more recent version, is required to walk through the exercises that we'll explore in this book. Check your hardware compatibility by comparing it with the system requirements provided on the Unity website at `https://docs.unity3d.com/Manual/system-requirements.html`.

As we'll be exploring AR development in this chapter, we will need either an Android or iOS device capable of supporting ARKit or ARCore. Review whether your device meets these requirements at `https://developers.google.com/ar/devices`.

Understanding the AR landscape

As we start our exploration of AR, it's crucial to first understand the foundational elements that enable this technology. How is it that our everyday devices can so effortlessly intertwine our physical reality with the digital? What mechanisms allow your device to sense, interpret, and interact with the world around it? And, perhaps most intriguingly, how can a simple screen transform into a doorway to an enhanced reality?

In this section, we aim to unpack the complex principles and mechanisms of AR, distilling them into a comprehensible format.

If terms such as AR, MR, and VR still seem opaque or interchangeable to you, consider revisiting *Chapter 1* for clarification. For now, our focus remains on AR, which transforms our world by superimposing it with elements of the digital domain. Let's look at the different types of experiences that AR offers.

What types of AR experiences exist?

The AR landscape is diverse, with experiences typically presenting themselves through one of several mediums: handheld mobile AR, AR glasses, or other types of AR such as projection-based AR or spatial AR. Each form has its unique characteristics, and their utilization depends on the context and the level of immersion desired. Let's learn more about these:

- **Handheld mobile AR**: Handheld mobile AR is perhaps the most widespread form of AR due to the ubiquity of smartphones. Imagine this type of AR as a window into an enriched reality. Through the screen of their smartphone or tablet, a user witnesses a mingling of the digital and the real. This overlay of digital content on a live camera feed breathes life into an otherwise static physical world. A prime example of this is the popular game *Pokémon Go*, where the user hunts for digital creatures that seem to inhabit our own world.

- **AR glasses**: AR glasses, on the other hand, provide a more immersive and hands-free AR experience. When the user dons these glasses, they step directly into an augmented world without the need for a separate device. Thanks to transparent displays, sensors, and cameras integrated into the glasses, digital information is seamlessly woven into the user's field of view. The implications of this technology are far-reaching, with potential applications in industries ranging from manufacturing and healthcare to entertainment. An example of a game designed for AR glasses is a new iteration of the Pokémon Go game, aptly named *Pokémon Go AR+*. Pokémon Go AR+ revolutionizes the original concept of the game, taking full advantage of AR glasses' capabilities. When wearing AR glasses, players are immersed in the Pokémon world more deeply. They can see Pokémon in their real-world surroundings as if they were actually there. Pokéstops and Gyms are visible in real-world locations, and players can interact with them directly. For example, if a Pokémon appears, the player can reach out to touch it and initiate a catch sequence. The game also allows for real-time battle simulations with other players using the AR environment.

> **Note**
>
> Though both handheld mobile AR and AR glasses act as pathways to AR, they differ significantly in their form factor, user experience, and level of immersion. Mobile AR serves as a portable gateway to an augmented world, accessible through a user's smartphone or tablet. On the other hand, AR glasses offer a fully immersive experience where the augmented world is in direct view. Given the prevalence of smartphones, this book will primarily focus on AR development for Android and iOS handheld devices.
>
> This does not mean, however, that AR glasses lack potential or usage. The focus on handheld devices mainly reflects their current wider use and accessibility. Despite this, AR glasses present significant opportunities and a level of immersion that handheld devices can't match. Even though the book focuses on handheld AR, the fundamental principles of AR development it covers, such as understanding the 3D space, user interaction, and user experience design, are largely applicable to AR glasses. Readers interested in AR glasses development can still gain valuable insights, though they might need to supplement their learning with additional resources specifically focused on AR glasses technologies.

- **Projection-based AR**: Projection-based AR casts digital imagery onto real-world surfaces. A notable application of projection-based AR is found in the automotive industry, where AR has started to make a significant impact. Information such as navigation, speed, and other essential data can be projected onto the vehicle's windshield, providing real-time visual cues to the driver without the need to take their eyes off the road.

- **Spatial AR**: Spatial AR uses holographic displays to create an illusion of virtual objects cohabitating in our physical environment. Holographic displays could be likened to digital mirages, using a combination of light projection and optics to create three-dimensional virtual objects that appear to float in space. This form of AR doesn't necessitate additional devices or wearables, enabling the user to interact with the holograms as if they were physically present.

Having explored the types of AR experiences, let's now delve into the techniques and concepts that enable the overlay of virtual objects in the real world.

What are marker-based and markerless AR?

AR is underpinned by a set of foundational principles that allow virtual objects to be accurately positioned and realistically interacted with in our physical world. These principles involve various technologies and techniques, such as **marker-based AR** and **markerless AR**.

Let's now learn more about each of these technologies.

Marker-based AR

Marker-based AR depends on specific markers or targets to initiate the display of AR content. These markers can take numerous forms, such as physical objects with identifiable patterns, QR codes, or images. When the AR system detects the marker, it overlays digital content onto it.

When developing AR applications, there are several types of markers that can be used to trigger the display of AR content. Here's an overview of some commonly utilized markers in AR:

- **Image markers**: These are distinct visual patterns or images that serve as the trigger for AR content. When an AR system detects these markers through the camera, it overlays the corresponding digital content onto them.

- **QR codes**: Quick response (QR) codes, which are two-dimensional barcodes containing specific information, can also serve as AR markers. When the camera or a QR code scanning library identifies these codes, it can trigger the display of certain AR content or interactions.

- **3D object markers**: These markers involve using specific physical objects as triggers for AR content. The AR system identifies the object's shape and features and uses this data to overlay the digital content.

- **Location-based markers**: These markers use the user's geolocation data to trigger AR content. By leveraging GPS or other location-tracking technologies, the AR system can overlay digital content that is relevant to the user's current location.

- **Code-based markers**: These are custom-designed patterns or symbols that can be recognized by the AR system and used as triggers. These markers are created and decoded using specific algorithms, providing a high degree of flexibility and customization for AR experiences.

The type of marker to use depends on the specific requirements of your AR application. Factors such as the desired user experience, tracking accuracy, and ease of marker recognition will guide marker selection.

While many AR experiences use markers, not all do, as you will learn in the next subsection.

Markerless AR

Markerless AR, also known as **position-based AR** or **simultaneous localization and mapping (SLAM)**-based AR, doesn't rely on predetermined markers or visual cues. It employs onboard sensors and complex algorithms to overlay digital information onto the physical world. Sensors can include **global positioning systems (GPS)**, accelerometers, and cameras to ascertain the user's location and orientation in the physical environment. With an understanding of the user's location, these AR systems overlay virtual content onto the physical surroundings based on geographical coordinates.

Let's delve into various options for implementing markerless AR and explore their real-world applications:

- **Geolocation-based AR**: This approach primarily uses GPS and is suitable for placing AR objects on a larger scale in outdoor environments.

  An example of this is Niantic's game Pokémon Go. Using the device's GPS, the game places virtual Pokémon creatures in real-world locations, allowing players to find and catch them.

- **Wi-Fi positioning system (WPS)**: This method determines the device's location based on the strength and origin of Wi-Fi signals. It's especially relevant for indoor AR experiences where GPS may be less effective. For instance, *Indoor Atlas* offers a platform for indoor navigation that combines magnetic information and Wi-Fi signals. This has been used to enhance AR experiences in shopping malls, guiding the user to specific stores or attractions with digital markers.

- **Bluetooth** and **ultra-wideband (UWB)**: These positioning techniques are geared for micro-location experiences, providing high precision in smaller spaces such as rooms or exhibits. A practical application can be seen in museums and galleries, where Bluetooth beacons are used in AR apps. These apps then serve multimedia content to visitors based on their proximity to specific artworks or exhibits.

- **SLAM**: SLAM is a more advanced technique that creates a digital map of the environment while tracking the user's location. This technique involves complex algorithms and uses the device's camera and other sensors. Imagine you're in a dark room and you light up a flashlight. As you move the flashlight around, you start to see and remember where different things are, such as chairs or tables. Over time, you build a map in your head of the whole room. SLAM does something similar. It's most suitable for applications that require accurate object placement and interaction in smaller spaces.

An example of SLAM is IKEA's app, *IKEA Place*. On the app, the user can select a piece of furniture from IKEA's catalog and the app overlays a 3D model of it onto the camera view, allowing the user to see how the item would look in their home.

- **Depth sensing**: This involves the use of advanced sensors such as **time-of-flight (ToF)** or **light detection and ranging (LIDAR)** sensors to capture the depth information of the surrounding environment. This method allows for more accurate placement and occlusion of virtual objects, where digital objects can correctly appear behind real-world objects based on depth information.

An example of this is Apple's *ARKit 4.0* platform, which incorporates the **Depth API** that leverages the LIDAR scanner available on some iPad and iPhone models. Depth API enables more realistic AR experiences. The *Complete Anatomy* app uses ARKit's depth sensing to place a detailed 3D human anatomy model into the real world, allowing the user to explore and interact with it as if it were physically present. Because of depth sensing, this model won't accidentally appear halfway inside your sofa but will stand correctly beside it.

- **Machine learning and AI**: Recent advances in AI and machine learning have opened new possibilities for markerless AR. Machine learning models can be trained to recognize different types of environments and objects, providing context for more intelligent and interactive AR experiences.

An example of this is Google's *ARCore* platform, which uses machine learning to recognize and augment specific objects or types of objects. Google's *AR Animals* feature uses ARCore to let the user search for an animal on Google and then view a 3D model of the animal in their space via AR.

Markerless AR offers immense possibilities for creating immersive and interactive AR experiences. Whether used in gaming, interior design, education, or a myriad of other applications, it has the potential to revolutionize how we interact with the digital world.

Understanding AR input types for interaction

Stepping into the realm of AR, one swiftly realizes that it's not just about the mesmerizing blend of physical and virtual realities that one can see. It's equally about how one can interact with these layered digital augmentations, a dimension defined by AR inputs. These inputs — modes by which the user interacts with the AR content — serve as a linchpin that shapes the overall AR experience.

As we venture further into this discussion, let's shine a spotlight on various AR input types and how they breathe life into real-world applications:

- **Touch input**: Touch input is a fundamental AR interaction. Simply put, the user can engage with the digital overlay through touch gestures on their AR device screen, be it a smartphone or a tablet. Touch input in AR includes not just tapping, but also other gestures such as swiping, pinching, and dragging. The specific gestures that can be used will depend on how the AR application is programmed. For example, a pinch gesture might be used to zoom in or out on an AR object, a swipe might rotate the object, and a drag could move the object around in the AR scene. The goal is to make the interaction with the AR elements as intuitive and natural as possible. *Snapchat lenses* provide a classic example of touch inputs at work. The user can animate the AR filters or induce changes by merely tapping different screen areas.

- **Device motion**: Device motion is another pivotal AR input. By harnessing the data from onboard accelerometers, gyroscopes, and magnetometers, AR applications can interpret the orientation and movement of the device as an input. This input type proves particularly useful for AR experiences that involve maneuvering through an environment or controlling virtual elements. A case in point is the game Pokémon Go, where the player can simulate the act of *throwing* Pokéballs by swinging their device.

- **Voice commands**: Voice commands infuse AR applications with hands-free and accessibility-friendly interaction. *Google Glass*, an AR eyewear device, employs voice commands as one of its core input methods. The user can simply say *Okay, Glass* followed by a command such as *get directions* or *take a picture* to interact with the device.

- **Eye tracking**: Eye tracking is typically employed in advanced AR glasses. By tracking the user's eye movements, these systems allow the user to interact with AR content just by looking at it. *North Focals* AR glasses are a good example of this technology in action. The user can steer a small, virtual cursor just by moving their eyes.

- **Hand tracking** and **gesture recognition**: Hand tracking and gesture recognition can be used in highly immersive AR systems to interpret and track hand movements, enabling the user to touch and interact with virtual objects directly. Microsoft's *HoloLens 2* is an example of this and allows the user to manipulate holograms with their hands, pinch to resize them, or tap them for interaction.

- **Physical controllers**: Physical controllers can range from handheld devices to wearable tech such as gloves, which provide tactile feedback and precise control in specific AR applications. For instance, the *Magic Leap One* AR headset is accompanied by a handheld controller, immersing the user into the AR experience by allowing them to interact with virtual content in a more nuanced manner.

As we have seen, AR inputs can significantly impact the immersive quotient of AR experiences. The choice of input methods hinges on the specific nature of the AR application. Therefore, understanding and implementing the most suitable input method can enhance an AR system's realism and usability manifold.

> **Note**
> Since this book primarily focuses on handheld mobile AR devices, to ensure that readers can easily reproduce all the projects presented within, we have chosen to restrict our scope to touch inputs.

In the next subsection, we will explore the AR toolkits available in Unity that can be utilized to implement the techniques we have just discussed.

Popular AR toolkits for Unity

In this section, we delve into Unity's suite of AR toolkits, providing an overview for both budding and experienced developers seeking to navigate through their AR development journey. Each toolkit offers unique capabilities that aid developers in crafting compelling AR experiences. Let's look at some of these now:

- **Vuforia**: Vuforia is a widely adopted AR platform, providing a blend of computer vision capabilities that accommodate both marker-based and markerless AR experiences. Its extensive feature set includes image tracking, object recognition, and target recognition, with wide-ranging platform support. An additional notable feature is Vuforia's cloud recognition, allowing developers to house a multitude of target images remotely, further expanding the AR experience's potential.

- **ARKit**: Within Apple's playground, ARKit is the optimal toolkit for experiences targeting iOS devices. It is crafted specifically to complement the iOS ecosystem, offering developers a suite of advanced features such as world tracking, face tracking, and scene understanding. These elements collectively serve to enrich the user's AR experience. ARKit predominantly uses Swift and Objective-C, Apple's proprietary programming languages. However, when integrated with Unity through the AR Foundation package, developers can leverage C#, enabling a more accessible and familiar programming environment.

- **ARCore**: Google's ARCore is the Android equivalent of ARKit, tailored for the world's most popular mobile operating system. ARCore equips developers with features such as environmental understanding, motion tracking, and light estimation, all of which are essential elements for crafting realistic AR experiences. Primarily, ARCore uses Java for native development. But, similar to ARKit's integration, ARCore can be incorporated into Unity projects via the AR Foundation package, allowing developers to use C#.

- **AR Foundation**: AR Foundation stands as Unity's high-level API package for constructing AR applications, unifying ARKit and ARCore's capabilities. This ingenious package enables a single, streamlined workflow for creating cross-platform AR experiences, eradicating the need to write separate code bases for iOS and Android. It's akin to owning a universal cookbook rather than individual recipe books for each cuisine. With AR Foundation, developers can leverage C#, a widely used, versatile programming language, making the process of crafting AR applications more efficient and intuitive.

If you were to use ARCore and ARKit separately, you would need to write separate sets of code for each platform. You would need to use ARCore to build your AR app on Android, and then rewrite the code using ARKit to make it work on iOS.

However, by using AR Foundation to develop AR apps, you only need to write a single set of code that works on both iOS and Android. AR Foundation provides a unified API that is compatible with both ARCore and ARKit, meaning you don't need to write different code for each platform. It simplifies the development process and saves a lot of time and effort.

> **Note**
> In this chapter, we'll focus on creating AR applications using AR Foundation.

However, it's important to remember that the use of ARCore or ARKit could offer unique advantages in certain scenarios, thanks to platform-specific features exclusive to each. For example, ARKit brings to the table LIDAR support from *version 3.5* onward. This feature utilizes the LIDAR scanner integrated into select iPhone and iPad models, offering refined scene understanding and precise depth estimation. Another ARKit exclusive is the *Motion Capture* feature from *ARKit 3* onward, enabling developers to record human movement and apply it to a 3D character model, effectively transforming the device into a motion capture studio.

On the other hand, ARCore also has a unique set of features for Android devices. One such feature is the Depth API, which generates depth maps using a single RGB camera. While ARKit also has depth sensing capabilities, ARCore's Depth API can function on a broader range of devices, even those without a dedicated depth sensor. Another distinctive feature of ARCore is *Augmented Images*, which allows the application to track and augment images at fixed locations, offering the potential for interaction with posters, murals, and similar items.

> **Important note**
> The examples mentioned earlier in this section may not be applicable when you are reading this book. Since ARKit, ARCore, and AR Foundation are constantly evolving, it is important to consult the latest documentation (`https://docs.unity3d.com/Packages/com.unity.xr.arfoundation@5.0/manual/index.html`) for the most up-to-date and accurate information.

Now, it is finally time to begin our first Unity AR project.

Setting up an AR project in Unity using AR Foundation

In this section, you will learn how to set up a simple AR project in Unity using AR Foundation. You will learn how you can place simple objects such as a cube, add plane detection functionalities, and implement touch inputs and anchors into your AR scenes.

Before using AR Foundation in our first AR application, however, we must first understand the architecture of this package.

Understanding AR Foundation's architecture

In this section, we'll delve into the exciting world of Unity's AR Foundation, a package that empowers you to create AR experiences across various platforms. Whether you aim to create applications for Android, Apple, or HoloLens, AR Foundation simplifies the process remarkably. Its extensive capabilities range from plane detection, image and object tracking, and face and body tracking, to point clouds and more. *Figure 4.1* breaks down the architecture of AR Foundation into a hierarchy of components that seamlessly work together to offer a consistent AR experience across different devices.

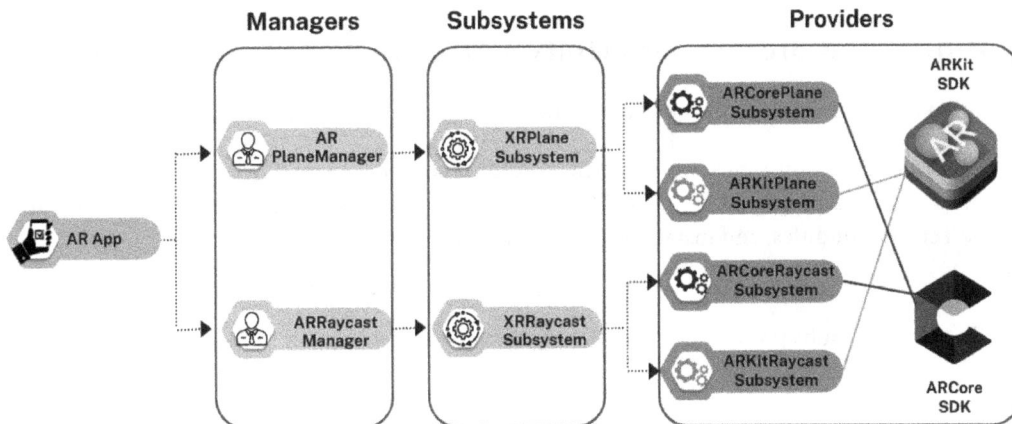

Figure 4.1 – AR Foundation's architecture

The **AR App** represents the application where developers craft their AR experiences. It is directly linked with the **Managers**, which serve as the primary interfaces for specific AR functionalities.

The **AR PlaneManager** is crucial for detecting and tracking real-world planes, such as floors or walls. It provides a consistent means to engage with the physical environment around a user, feeding the app with data about surfaces in the user's surroundings.

Similarly, the **ARRaycast Manager** plays an essential role in understanding user interactions within this AR space. It casts rays into the AR scene, determining where these rays intersect with real-world surfaces. This is particularly vital when users intend to place virtual objects on these surfaces or wish to interact with virtual elements in relation to the real world.

Delving deeper, these managers interact with **Subsystems**, abstract layers that communicate with the actual platform-specific modules. The **XRPlane Subsystem** standardizes data related to plane detection, ensuring that plane-related events and data are uniform, irrespective of whether it's ARKit or ARCore doing the underlying work. The **XRRaycast Subsystem** offers a consistent interface for raycasting, abstracting the nuances of each platform's approach.

Lastly, the **Providers** represent the platform-specific SDKs that power AR on each device type. ARKit SDK is Apple's contribution for its iOS devices, with specialized subsystems for plane detection and raycasting. On the other side, ARCore SDK is Google's solution for Android, mirroring the functionalities offered by ARKit but tailored for the Android ecosystem.

As you can see, AR Foundation offers XR developers a unified framework by allowing them to focus on their application's AR experience without worrying about platform-specific intricacies. Through its layered architecture, it ensures that applications remain consistent and high-quality, whether deployed on an iPhone or an Android device.

With this understanding, it is finally time to create our first AR project in Unity.

Creating an AR project with Unity's AR template

Creating an AR project in Unity is quite straightforward. Follow these steps to get started:

1. Depending on your target AR platform — Android or iOS — you'll need to navigate first to the `Installs` folder in the Unity Hub, click on the **Settings** icon adjacent to your version, select **Add Modules**, and install the appropriate build support.

    To show your project in the AR Unity template, you'd follow the subsequent steps.

2. In the Unity Hub's project window, click on the **New project** button in the top-right corner.

3. Choose the **AR** template and give your project a unique name of your choice, as shown in *Figure 4.2*.

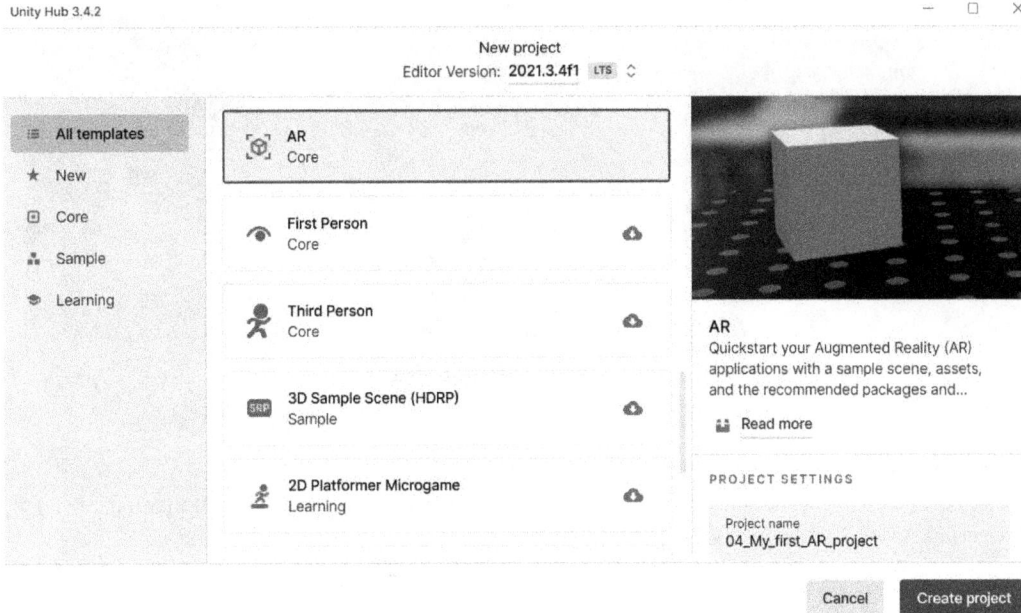

Figure 4.2 – The AR template available in the Unity Hub

> **Why choose the AR template?**
>
> In Unity's world, the AR template is like a bountiful garden, already seeded with pre-installed packages such as AR foundation, ARKit Face Tracking, ARKit XR-, ARCore XR-, Magic Leap XR-, and the OpenXR plugins, not to mention a sample scene. However, you might observe that this lush garden contains plants that aren't necessary for your target landscape: Android or iOS. To maintain a cleaner garden, you can selectively pick the seeds you need – the OpenXR plugin, AR Foundation, Input System and the ARCore XR plugin (for Android) or ARKit XR plugin (for iOS) – from Unity's own nursery, the package manager, and plant them in a regular 3D scene.

4. Finally, click on **Create Project**.

Once the scene loads in the Unity Editor, it should look like what is shown in *Figure 4.3*.

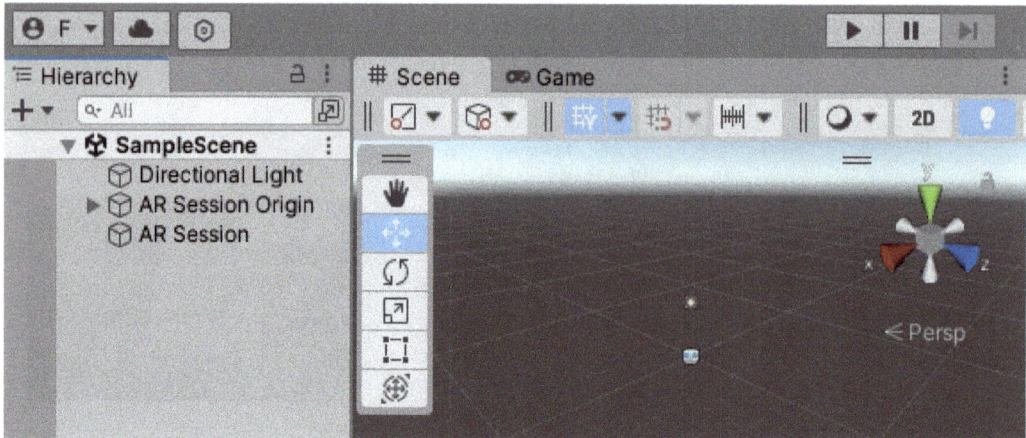

Figure 4.3 – Unity's AR SampleScene

> **Important note**
>
> If you don't see the scene shown in *Figure 4.3* in the Unity Editor, you can open it manually by selecting **Assets | ExampleAssets | SampleScene**.

The newly opend scene contains **Directional Light**, **AR Session Origin**, and **AR Session** GameObjects. If you wish to bring these objects into a new scene, you can conjure an **AR Session** GameObject by right-clicking in the hierarchy and choosing **XR | AR Session**. The **AR Session Origin** GameObject can be summoned in a similar fashion by selecting **XR | AR Session Origin**. That's the initial magic performed.

> **Important note**
>
> At the time of this book's publication, **AR Session Origin** and **AR Session** represent the default GameObjects integrated into the *AR Foundation*'s template. Yet, given the dynamic nature of *AR Foundation*, continuous updates and revisions are anticipated. There are indications that **AR Session Origin** is slated to evolve into *XR Origin* (*Mobile AR*) in upcoming versions.
>
> Should you encounter such changes, don't fret. The essence of this chapter remains intact and can be easily navigated. We recommend accessing our *GitHub* repository to clone the specific project versions we've worked with, ensuring compatibility with Unity. Often, transitioning from outdated GameObjects to their newer counterparts is seamless, requiring minimal, if any, adjustments to the existing logic.
>
> Stay adaptable and remember: the core concepts and foundations presented here remain your guide, even as the tools evolve.

Before we start inspecting these GameObjects, we first need to adjust some project settings of our AR scene.

Changing the project settings

To define which providers we want to target with our AR scene, go to **XR Plug-in Management** under **Edit | Project Settings**. Depending on the platform we're targeting, be it an Android or an iOS device, we activate the corresponding ARCore or ARKit checkbox. If you seek to embrace both platforms, enable both checkboxes. This will install the provider packages into your project.

Important note

If you don't see Android, iOS, or both tabs, it's likely that the *Android* or *iOS Build Support* module hasn't been integrated into your Editor. Let's see how we can fix this problem.

Figure 4.4 shows the installed Unity Editor with the Build Support modules marked in the Unity Hub.

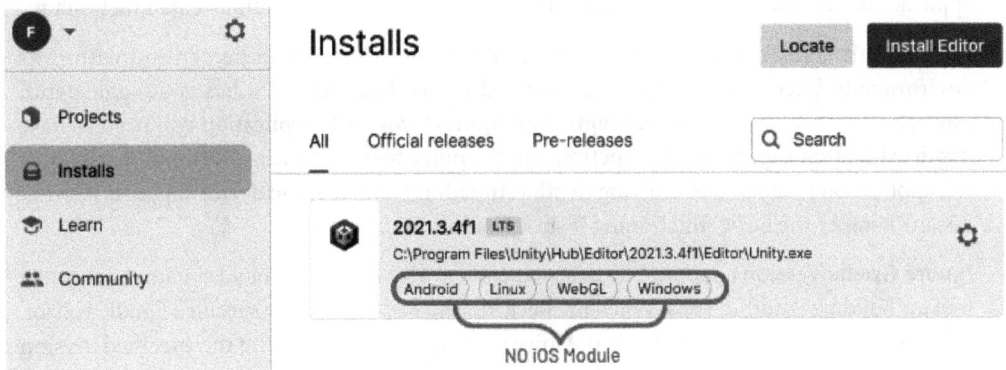

Figure 4.4 – The installed Unity Editor with the Build Support modules marked

The displayed image indicates an absence of the *iOS Build Support* module in the Unity Editor. To adjust this, go to the **Installs** tab in the Unity Hub. Click the **Settings** icon in the top-right corner of your Unity Editor installation. Then, select **Add Modules**, enable the checkboxes for **iOS Build Support** or **Android Build Support** – or both, if you will – and install them. Once the installation is finished, go back to your project's **XR Plug-in Management** window.

Now, you should find the **ARCore** and **ARKit** tabs under **XR Plug-in Management**.

Now, select the **ARCore** tab to explore fields related to the configuration and settings of the ARCore packages in your Unity project. *Figure 4.5* shows the **ARCore** tab with its default settings.

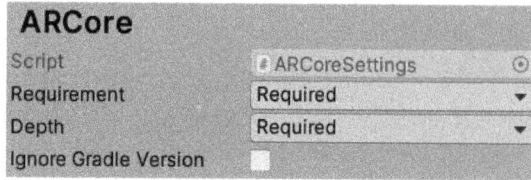

Figure 4.5 – The ARCore tab with its default settings

Let's go through the function of each field:

- **Requirement**: This parameter dictates the necessity of ARCore for your application. If set to **Required**, it establishes ARCore as an integral component for your target device. Conversely, when marked as **Optional**, ARCore is positioned as an additional feature, supplementing the application's capabilities when present but not essential for the application's core functionality.

- **Depth**: This is the tool that gives your device the power of depth perception within its environment. Useful for occlusion and other advanced AR effects, it lets you toggle depth estimation in ARCore. If you set **Depth** to **Required**, your AR application will need to have depth estimation capabilities to function. On the other hand, selecting **Optional** means your AR application is capable of utilizing depth estimation features if the device supports it, but it doesn't hamper the basic functioning if such features are absent.

- **Ignore Gradle Version** checkbox: This is your control over the Gradle build system, an essential tool for building Android apps. When unchecked, Unity adheres to the specified Gradle version for your project. However, if checked, Unity gains autonomy, ignoring the specified version and choosing to operate on its default version.

While the default settings – having both the **Requirement** and **Depth** fields set to **Required** and the **Ignore Gradle Version** checkbox unchecked – suit most projects, there are exceptions. For instance, if you're creating an AR application designed for a wide variety of devices with varying capabilities, or if your application's main functionalities do not heavily rely on ARCore's depth perception, you might want to set the **Depth** field to **Optional**. Furthermore, if you're working on a project that requires a specific Gradle version or has conflicts with newer versions, checking the **Ignore Gradle Version** box would be necessary. However, for this chapter, we can keep the default settings.

Next, we turn our attention to the **ARKit** tab. Just like its ARCore counterpart, these fields are connected to the configuration and settings of the ARKit packages in your Unity project. *Figure 4.6* shows the **ARKit** tab with its default settings.

Figure 4.6 – The ARKit tab with its default settings

Let's decode the roles of all the fields shown in *Figure 4.6*:

- **Requirement**: Acting in a manner similar to ARCore's **Requirement** field, this indicator announces whether ARKit is deemed vital for your project.

- **Face Tracking**: Acting as a switch, this checkbox enables or disables the face tracking functionality in ARKit. When activated, it allows tracking and recognizing a range of facial features and expressions using the device's front-facing camera.

For the moment, we'll abide by the default settings – the **Requirement** field standing firm at **Required** and the **Face Tracking** checkbox remaining in the off position, as it's not currently needed.

In the forthcoming sections, we're going to delve into the GameObjects that accompany the **AR** template. We'll dive into their roles and demystify why they hold such significance. Let's begin with the AR Session GameObject.

Understanding the AR Session GameObject

The **AR Session** GameObject manages the lifecycle of your AR application. To understand it, click on it in the scene hierarchy and navigate to the inspector window. As shown in *Figure 4.7*, the **AR Session** GameObject contains the **AR Session** and the **AR Input Manager** components.

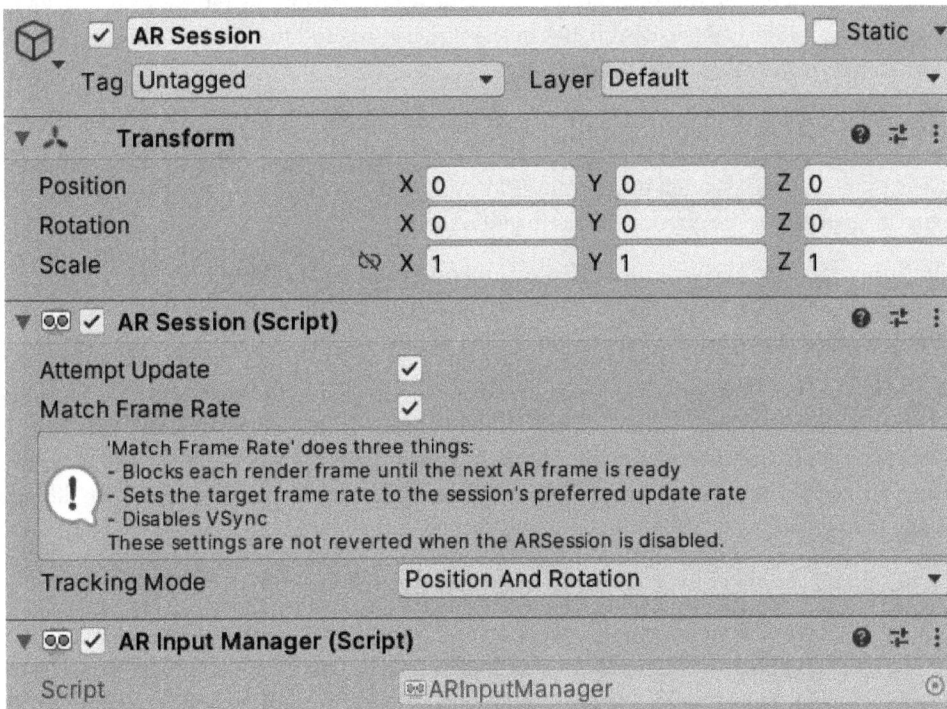

Figure 4.7 – The Inspector window when the AR Session GameObject is selected

Let's understand each of the fields shown in *Figure 4.7*:

- **Attempt Update** checkbox: This checkbox, when marked, is your instruction to the AR session to be vigilant about keeping the device's pose and tracking state up-to-date with every frame and every tick of the clock. Checking the box commands the AR session to fetch fresh tracking data diligently, ensuring that even when the device's tracking wavers temporarily, the AR elements maintain their alignment with the real world. Imagine an AR hoop shooting game. In this game, you see a virtual hoop overlaid onto your real-world environment through your device's screen. The objective is to shoot virtual balls into this hoop from various angles and distances. For this to function well, it's vital that the tracking data of your device is continuously updated. That's where the **Attempt Update** checkbox comes in.

  Activating **Attempt Update** commands the game to consistently refresh the tracking state, securing the harmony of the virtual objects with the real-world environment, notwithstanding transient tracking losses.

- **Match Frame Rate** checkbox: This, when marked, means that you're asking the AR session to walk in step with the device's camera frame rate. The AR session adjusts its rhythm to mirror the camera's, culminating in a harmonious visual experience. Consider an AR app that lets the user envision virtual furniture within their actual space.

  A ticked **Match Frame Rate** checkbox ensures the AR session's frame rate keeps time with the camera's, curbing visual discord between the actual space and the virtual furniture, thereby enabling a precise assessment of aesthetics.

- **Tracking Mode** drop-down menu: This is the key to defining the tracking quality of your AR session. The selected mode determines the AR system's awareness of, and response to, the device's motion and position in the real world. Picture an AR navigation app that superimposes virtual arrows over the physical world to guide the user.

  In this navigation app, opting for **Rotation and Position Tracking** imparts full tracking capabilities, ensuring the virtual arrows trace the contours of the real world accurately, guiding the user along their path with precision.

For our current exploration, we'll stick with the default settings – both the **Attempt Update** and **Match Frame Rate** checkboxes are ticked and **Tracking Mode** is set to **Position And Rotation**. In the future, your choice of these settings should reflect your project's goals and the intended user experience. To discover the best blend for your needs, we encourage you to experiment with different settings and configurations.

The second component under the AR Session umbrella is the **AR Input Manager** component.

The AR Input Manager component translates the user's interactions with the AR scene into meaningful input. It perceives and processes a variety of user inputs such as taps, touches, and gestures. It's the invisible hand enabling interactivity in your AR application, allowing placement of objects or interactions with virtual elements. For instance, imagine an AR game where the user neutralizes virtual targets with a tap. The **AR Input Manager script** reads the tap on the screen and furnishes the requisite information to unleash the shooting action in the game.

Now, our focus shifts to the AR Session Origin GameObject.

Exploring the AR Session Origin GameObject

The **AR Session Origin** GameObject is the primary element controlling the synchronization between the physical world and virtual objects projected in AR. It comprises various components, such as AR Session Origin, AR Plane Manager, AR Anchor Manager, AR Raycast Manager, and Anchor Creator, as we can see in *Figure 4.8*.

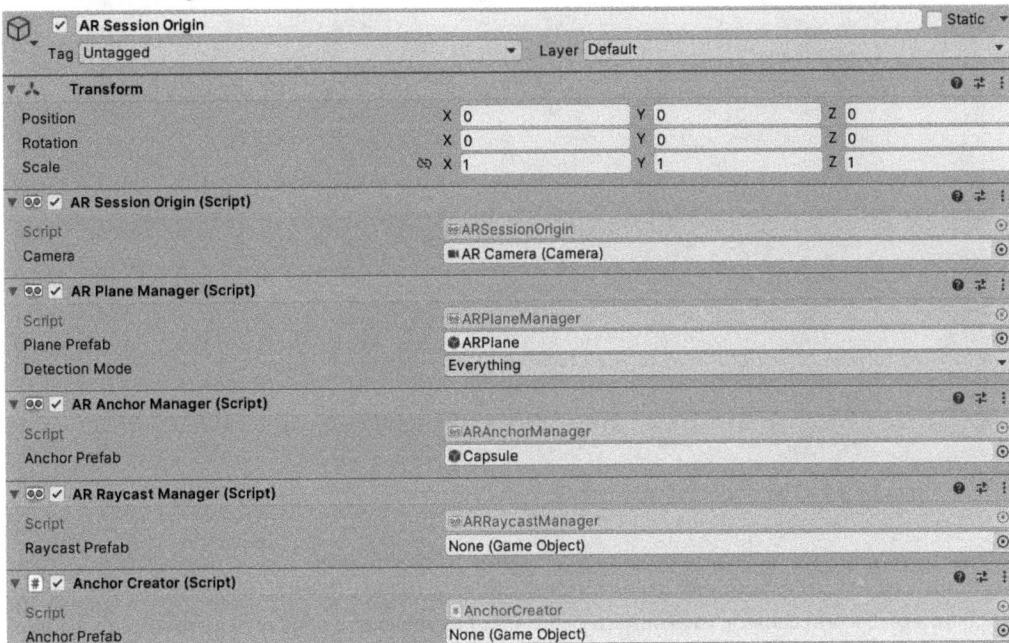

Figure 4.8 – The Inspector when AR Session Origin is selected

For the initiation phase of our AR development, we only require the AR Session Origin component. Hence, you may disable all other components for the time being.

Think of AR Session Origin as the pivot point of your AR experience. It provides a basis for anchoring virtual objects within the AR environment. Let's visualize an AR app allowing the user to place virtual furniture within their home. Here, the AR Session Origin component arranges and alignes the virtual furniture correctly within the user's physical space.

AR Session Origin ensures the alignment of the user's environment with the position and orientation of the displayed AR scene. For instance, when a user places a virtual chair, this script maintains the chair's positioning and orientation, no matter the user's movement within the room. This is why this component is essential in creating an authentic AR experience.

For blending the virtual objects with the real-world footage, the AR session necessitates the presence of a camera. In the case of an existing main camera in your scene, you can safely remove it. AR Session Origin incorporates its own child **AR Camera**. This camera, devoid of any skybox for a monochromatic background, comes pre-equipped with necessary components, such as **Camera**, **Tracked Pose Driver**, **AR Camera Manager**, and **AR Camera Background**, as shown in *Figure 4.9*.

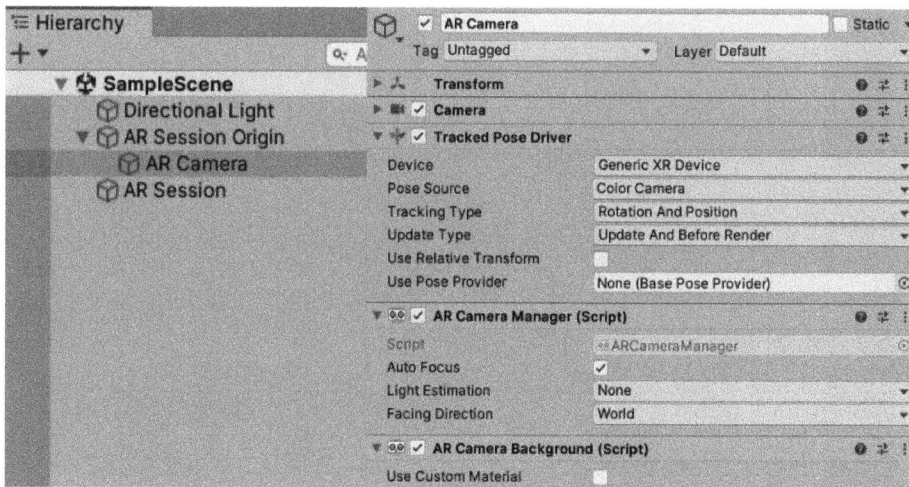

Figure 4.9 – The Inspector when AR Session Origin's child AR Camera is selected

Let's now learn more about each of these:

- Like a real-world camera, the **Camera** component captures and displays a portion of the game scene to the player – essentially, what you see on your device's screen when playing an AR game. It can be positioned and rotated to capture different views within the game scene.

- The **AR Camera Manager** component controls the phone camera settings, such as **Facing Direction**, **Light Estimation**, and **Auto Focus**. **Auto Focus** makes the camera automatically adjust its focus to keep AR objects sharp, just like autofocus in a regular camera. **Light Estimation** allows you to choose how the AR system estimates real-world lighting conditions to make virtual objects look more realistic. Options range from not estimating lighting at all (**None**) to estimating various aspects of lighting, such as **Ambient Light Intensity** and **Color**, and the main light source's **Direction** and **Intensity**. **Facing Direction** decides the direction in which the AR Camera is facing. The **World** setting is typically for the rear-facing camera (seeing the environment), and **User** is for the front-facing camera (selfie mode). **None** disables this setting. For now, we can leave these settings as they are.

- The **Tracked Pose Driver** component takes information about the device's physical position and orientation, collectively referred to as its pose, and uses that to set the position, rotation, and scale of the camera within the Unity game scene. This is what's referred to as the camera's **Transform**. This ensures that the AR scene remains in harmony with the phone's movements. To facilitate this, the Tracked Pose Driver provides several configurable options. **Device** lets you select the type of device being tracked. **Pose Source** allows you to choose the part of the device that provides the tracking data. The **Tracking Type** drop-down menu indicates what type of movement (rotation, position, or both) is tracked.

  Update Type dictates when the tracking information updates, whether during the regular update cycle, before rendering the next frame, or both. If the **Use Relative Transform** checkbox is ticked, the tracking data is based on the device's position and rotation relative to its initial state.

  Finally, the **Use Pose Provider** field can be enabled to utilize an external source for tracking data, which is potentially useful for specialized tracking systems. Together, these settings give **Tracked Pose Driver** the flexibility to handle a wide range of AR scenarios.

- The **AR Camera Background** component controls how the real-world view captured by your device's camera is displayed in your AR scene. The **Use Custom Material** checkbox allows you to apply a custom material, for example, to add special visual effects to the real-world view. If left unchecked, the real-world view is displayed as is, without any additional effects.

The AR Session Origin GameObject performs the critical task of converting AR session space into Unity's world space. Given the unique coordinate system employed in AR (referred to as AR session space), this conversion is vital for accurately positioning the GameObjects relative to the AR Camera.

With this information about the most important GameObjects of our AR scene in mind, it is time to place a simple object in our AR scene.

Placing a simple cube into the AR scene

To begin testing our AR functionality, we need a 3D object in our scene to visualize the AR effect. For this purpose, you can create a simple cube object. Right-click in your project hierarchy and choose **3D Object | Cube**. Position this cube at coordinates of (0,0,3) and apply a rotation of (30,0,0), as shown in *Figure 4.10*.

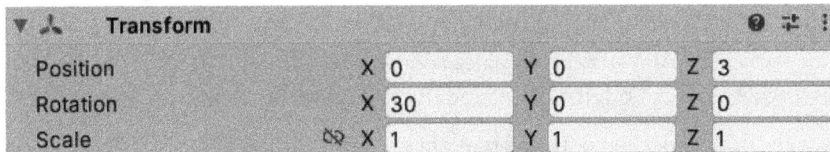

▼ 人 Transform						❷ ⇄ ⋮
Position	X	0	Y	0	Z	3
Rotation	X	30	Y	0	Z	0
Scale	X	1	Y	1	Z	1

Figure 4.10 – The Transform values of the cube

These specific coordinates and rotations are only examples and can be altered according to your requirements.

The key objective is to position the object somewhere in the direction of the AR Camera and at a distinct distance. This placement will help verify whether the AR system is correctly overlaying virtual content onto the real world, as it provides a benchmark for alignment with your physical surroundings. It also allows you to gauge the depth perception capabilities of your AR system.

To validate whether the cube will be correctly displayed in the AR view, you can either select the AR Camera in the hierarchy – this will cause a camera view to appear in the bottom-right corner of the **Scene** window – or switch to the **Game** window, as shown in *Figure 4.11*.

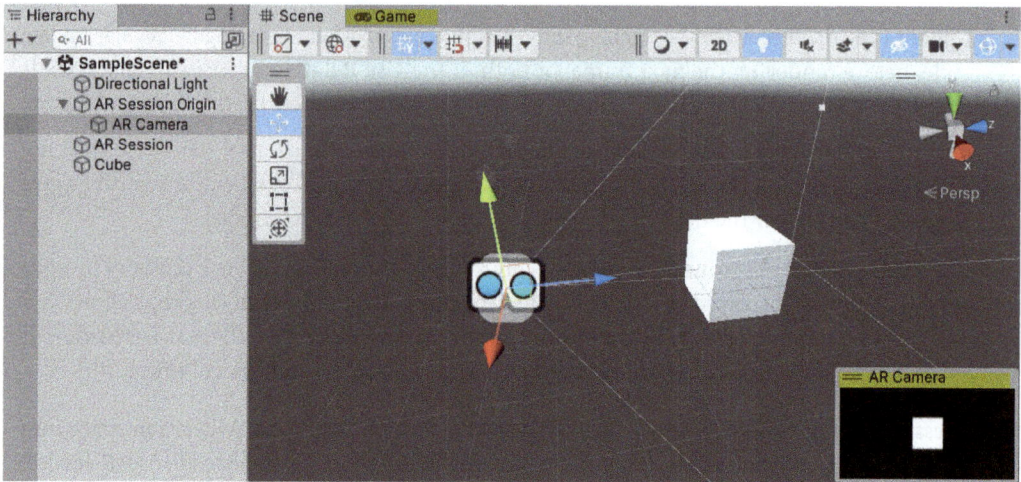

Figure 4.11 – The Unity project showing how to validate whether
the cube will be correctly displayed in the AR view

In an AR application, AR Session Origin typically corresponds to the starting location of your device's camera when the AR experience begins. However, positioning a virtual object directly at **Session Origin** can indeed cause some issues. Here are a few potential problems that could arise:

- **Inaccurate scaling**: If you were to place a virtual cube directly at **Session Origin**, it could appear disproportionately large or small compared to real-world objects. It's like holding a real-life object very close to your eyes – even a small object can appear large.

- **Interference with tracking**: **Session Origin** serves as a reference point for tracking the device's movements in the real world. If you place a virtual object at this exact point, the AR system might struggle to track both the object and the device's movements accurately.

- **Unpleasant user experience**: If a virtual object is positioned directly at **Session Origin**, it might appear too close to the user or even obstruct the user's view of the rest of the AR scene, creating a less enjoyable experience.

> **Tip**
>
> To avoid these issues, it's generally recommended to position virtual objects at a reasonable distance from **Session Origin**. The precise location would depend on the specific requirements of your AR experience. For instance, if you're creating an AR furniture placement app, you might position virtual furniture based on the detected real-world surfaces, such as floors and tables. By doing so, you ensure that the virtual furniture appears at an appropriate scale, doesn't interfere with tracking, and provides a pleasant user experience by appearing in the correct place in the real world.

After placing your cube, you can proceed with testing the scene by following the instructions outlined in the *Deploying AR experiences onto mobile devices* section of this chapter. As it stands, your AR application is simple, featuring a static object positioned in front of the user. However, we're aiming for a more interactive experience, so we'll delve into the anchor and plane detection functionalities in the upcoming section.

Implementing plane detection in AR Foundation

In this section, we will make use of AR Foundation's **plane detection** capabilities. AR Foundation's plane detection forms the crux of numerous practical applications by providing a cornerstone understanding of real-world environments. Its ability to identify and comprehend flat surfaces enables digital objects to interact meaningfully and realistically with physical spaces, enhancing the overall augmented reality experience.

In the retail and e-commerce sectors, plane detection is fundamental to accurate product visualization. It ensures a virtual couch is placed right in front of your living room wall and not in a random position in your room.

This technology also underpins the immersive experiences in AR gaming. By identifying real-world surfaces, games such as Pokémon Go can accurately place virtual elements, enhancing the thrill of the chase. Other games use this technology to create worlds on your coffee table, maintaining the illusion by ensuring that virtual objects interact convincingly with the physical planes.

To make use of plane detection, enable the **AR Plane Manager** component you disabled in the *Exploring the AR Session Origin GameObject* section. Then, you can also delete the cube as we don't need it anymore. Now, select **AR Session Origin** and inspect the **AR Plane Manager** component in the hierarchy.

The **AR Plane Manager** component in Unity's AR Foundation is responsible for detecting and tracking real-world surfaces, or planes, via your device's camera. It utilizes AR technology to understand the physical environment, creating digital representations that can interact with virtual objects.

Here's a breakdown of what **AR Plane Manager** does:

- **Plane Prefab**: This is essentially the digital template or blueprint used to represent detected planes. When **AR Plane Manager** identifies a flat surface (a plane) in the real world, it creates a **Plane Prefab** instance in the AR space. Each prefab instance corresponds to a specific real-world plane, providing a surface on which you can place or interact with virtual objects.

  Let's say, for example, you're creating an AR game where virtual cats roam around. The **Plane Prefab** instance could be a flat surface that the cats can walk on. If your living room floor is detected as a plane, a **Plane Prefab** instance will be created, providing a surface for your cats to frolic on.

- **Detection Mode**: This setting determines the orientation of planes that **AR Plane Manager** should detect. Here are some components within **Detection Mode** that would be useful for you:

  - **Everything**: This setting combines the detection of horizontal and vertical planes by **AR Plane Manager**. If you're creating an AR furniture placement app, you might use this setting. The app could then detect both the floor (a horizontal plane, for placing a chair) and the walls (vertical planes, for hanging pictures).

  - **Horizontal**: **AR Plane Manager** will detect only flat surfaces oriented horizontally, such as floors and tabletops. For instance, if you're creating an AR game where characters run around on the floor, you'd likely use this mode.

  - **Vertical**: Conversely, this setting will detect only vertically oriented surfaces, such as walls and doors. This might be used in an AR interior design app that allows the user to place virtual paintings or wallpapers on their walls.

As we want to detect horizontal and vertical planes, the drop-down selection can remain at **Everything**.

Next, you will notice that the AR template already has an **ARPlane** prefab assigned. This prefab is what you will see on top of the detected surfaces. As this prefab is already configured, you can use the **ARPlane** prefab. However, you can also create your custom-made plane prefab. This can be done by right-clicking in the hierarchy window and selecting **XR | AR Default Plane.**

AR Default Plane GameObject in Unity's AR Foundation is composed of various components, each of which performs a specific function within the overall system. Here's a breakdown of what each of these components does:

- **AR Plane**: This script controls the essential behavior of a plane in the AR environment. Two notable fields in this script are as follows:

  - **Destroy on Removal** checkbox: If checked, when a detected plane is removed or no longer needed, the corresponding **Plane** GameObject is destroyed or removed from the Unity scene to free up resources. Imagine this to be like cleaning up toys when you're done playing with them to make space for other activities.

- · **Vertex Changed Threshold**: This field controls how sensitive the plane is to changes in its detected shape. If the real-world surface changes, the plane's mesh vertices will need to be updated. This threshold determines how much change is needed before an update occurs. It's like deciding when to adjust a puzzle because the pieces have moved – too often can be unnecessary, yet too seldom and the picture may not make sense.

- **AR Plane Mesh Visualizer**: This script is responsible for rendering the visual representation of a detected plane. Imagine you're using an AR app on your phone to preview a piece of furniture in your room before buying it. When you point your phone's camera at the floor, the app detects a flat surface – a plane. **AR Plane Mesh Visualizer** then creates a visible grid or pattern overlay on your phone screen that represents this detected plane.

- **Tracking State Visibility** dropdown (**None, Limited, Tracking**): This setting determines when the plane should be visible based on the tracking quality. For instance, you might only want to see the plane when it's fully tracked (good quality), or also when tracking is limited (lower quality). It's like deciding when to display a picture – if it's blurry, you might prefer to wait for it to load until it's clear.

- **Hide Subsumed** checkbox: If checked, when one detected plane is subsumed by another (i.e., entirely overlapped by a larger plane), the smaller plane is hidden. This can make the AR scene less cluttered and more efficient. Think of this to be like removing smaller rugs when you put down a larger one that covers them completely.

- **Mesh Collider**: This component allows virtual objects to physically interact with the plane as if it were a solid surface. For example, if you drop a virtual ball onto the plane, **Mesh Collider** ensures that the ball bounces back rather than falling through the surface.

- **Mesh Filter** and **Mesh Renderer**: These work together to create and display the 3D mesh for the plane. **Mesh Filter** generates the shape of the plane (the mesh), while **Mesh Renderer** applies materials and textures and renders it on the screen. They're like a sculptor and a painter respectively, working together to create a lifelike statue.

- **Line Renderer**: This component is often used to draw the border around the detected plane, helping the user to see the extent of the plane. It's like drawing a chalk outline around an area to indicate its boundaries.

All these components work in harmony to detect, represent, and interact with real-world surfaces in your AR applications, ensuring a seamless blend of virtual objects with the physical environment. You can keep the default settings for our scene.

After creating **AR Default Plane GameObject**, you can proceed to turn it into a prefab. To start, right-click in **Project Window** and choose **Create | Folder**. Name this new folder Prefabs. Next, drag **AR Default Plane GameObject** into the Prefabs folder. Doing so will automatically create a prefab out of it. With **AR Session Origin** now selected, drag the **AR Default Plane** prefab to the appropriate **Plane Prefab** field of the **AR Plane Manager** component within the Inspector, as illustrated in *Figure 4.12*.

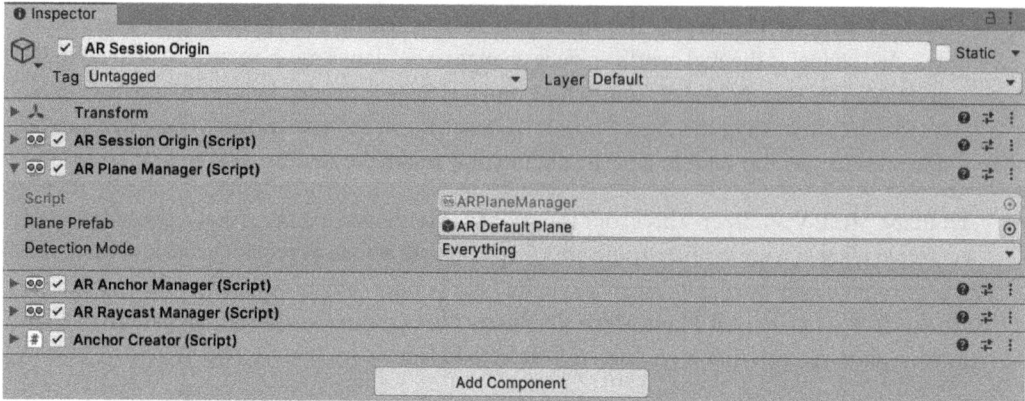

Figure 4.12 – The AR Default Plane prefab assigned to the appropriate
Plane Prefab field of the AR Plane Manager component

Now, your application would be able to detect a plane in your environment and place the plane prefab of your choice on top of it.

Figure 4.13 shows the default plane prefab of Unity's AR template.

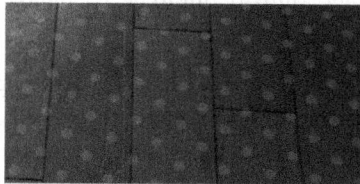

Figure 4.13 – The plane prefab of Unity's AR template detecting the wooden floor of an apartment

As we can see, with AR Foundation, it is straightforward to incorporate plane detection into a project and to project patterns onto detected surfaces. In the upcoming section, we will merge these concepts with the use of anchors and touch inputs. This combination will enable us to place an object onto a detected surface simply by tapping on the screen.

Implementing touch inputs and anchors

To get started on working with touch inputs and anchors, enable the **AR Anchor Manager**, **AR Raycast Manager**, and **Anchor Creator** components in the Inspector when **AR Session Origin** is selected. These components are all we need to get the touch input and anchors to work. Let's break each of them down:

- **AR Anchor Manager** script: **AR Anchor Manager** operates as the custodian of object stability in your AR space. When you designate a position for a virtual object within your AR scene, **Anchor Manager** ensures the object remains tethered to that specific real-world location, irrespective of changes in device perspective. It is akin to an overseer, maintaining each virtual object in its correct placement relative to the real-world coordinates.

- **AR Raycast Manager** script: Acting as the tactile sense of your AR application, **AR Raycast Manager** emits invisible rays from your device into the scene. When these rays encounter a surface, they return data about its position and orientation. This process is key to interactions such as placing a virtual object on a real-world surface, as it enables your application to perceive and understand the topography of the environment.

- **Anchor Creator** script: This script functions as an efficient tool, facilitating the creation and placement of new anchors within your scene. Given that anchors are pivotal to securing virtual objects in a consistent real-world location, **Anchor Creator** simplifies the process of generating these anchor points. This can range from situating a new virtual object to immobilizing an object in transit.

Even though all components have a **Prefab** field, we just need to assign a prefab to the **AR Anchor Manager** component. This prefab will be spawned when we tap on the screen.

When you tap on the screen, **AR Raycast Manager** determines where in the real world you've indicated by casting a ray from your screen touch point into the AR scene. If this ray hits a detected surface, **AR Anchor Manager** creates an anchor at this real-world location and ties the prefab to it. This process ensures the stable positioning of your virtual object in the AR scene.

The prefab attached to **AR Anchor Manager** essentially acts as a template for the object you want to instantiate (or create) in the AR environment whenever the screen is tapped. This is why it's essential to assign a prefab to **AR Anchor Manager**.

The other components do not require a prefab because they are not directly responsible for creating objects in your AR scene. **AR Raycast Manager** deals with detecting the real-world surfaces you are interacting with, and **Anchor Creator** facilitates the creation and placement of the anchors themselves. Neither of these tasks necessitates the creation of a new object from a prefab.

For demonstration purposes, we use a simple capsule primitive. Simply right-click in the hierarchy and select **3D Object** | **Capsule**. Downscale the capsule to (0.2,0.2,0.2). Now, create a new folder in the `Project` folder called `Prefabs` (right-click + **Create** | **Folder**) and drag the capsule into this folder. This will automatically create a prefab of the capsule that you can drag into the **Anchor Prefab** field of the **AR Anchor Manager** component.

Now, your app should spawn the capsule on a touch input. *Figure 4.14* shows the deployed application.

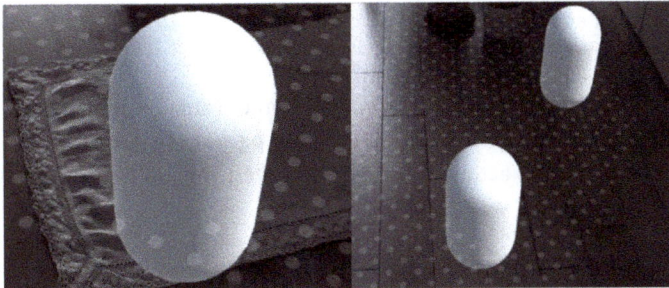

Figure 4.14 – The deployed application

The preceding image illustrates the instantiated capsule prefabs following screen taps. These capsules are positioned atop the detected surface, confirming the effective operation of the AR systems. Specifically, **AR Raycast Manager** accurately casts rays from the screen touch point onto the detected surface, and in response, **AR Anchor Manager** successfully creates anchors, anchoring the prefabs at the indicated locations. Everything should now be working perfectly, so it's time to build the scene onto your device. Depending on whether you're using an Android or iOS device, you can navigate to the relevant subsection in the following section.

Testing AR experiences directly in Unity

As of *AR Foundation 5.0*, developers can conveniently test AR scenes right in the Unity Editor using the **XR Simulation** feature, without the constant need to deploy on mobile devices. By the time you read this book, this feature might already be pre-installed. Let's quickly check that by following these steps:

1. Navigate to **Edit | Project Settings**.

2. Choose **XR Plug-in Management**.

3. Look for the **XR Simulation** option under **Plug-in Providers** and enable it.

If you don't see the **XR Simulation** option, you'll need to manually install *AR Foundation 5.0* or a newer version. The next section explains how to do this.

Installing AR Foundation 5.0 or later versions and related packages

When you edit your project manifest, you control which package versions Unity loads into your project. There are two ways to edit your project manifest: add a package by name in **Package Manager**, or manually edit the project manifest file. Let's do it in **Package Manager**:

1. Select **Window | Package Manager** to open the **Package Manager** window.

2. Click on the small + icon in the top-left corner and select **Add package by name**. Type in `com.unity.xr.arfoundation`. This will automatically add the most recent version available. At the time of writing, this is version **5.1.0**.

   If you want to use a specific version, you can type in your desired version in the **Version (optional)** field.

3. Continue by updating the **OpenXR** plugin. Click on the small + icon located in the top-left corner of the **Package Manager** window, select **Add package by name**, and then input `com.unity.xr.openxr`. This action will import the latest version of the **OpenXR** plugin.

4. If you're upgrading from *AR Foundation 4* to a newer version, uninstall both the **com.unity.xr.arkit-face-tracking** and **com.unity.xr.arsubsystems** packages if they are currently in your project. To do this, open **Package Manager** and set the **Package** dropdown to **In Project**. Utilize the search bar located at the top-right corner to look for com.unity.xr.arkit-face-tracking and com.unity.xr.arsubsystems. If they appear in the search results, proceed with uninstallation; otherwise, they are not present in your project.

5. Next, depending on your targeted mobile device platform, it's important to update either the *ARCore* or *ARKit* packages. Navigate to the **Add package by name** option within the **Package Manager** window. For Android devices, input com.unity.xr.arcore and ensure its version aligns with that of *AR Foundation*. For iOS platforms, input com.unity.xr.arkit, making sure its version matches *AR Foundation's*.

> **Note**
>
> Always remember to refer to the *AR Foundation* documentation for the most recent version details and Editor compatibility: https://docs.unity3d.com/Packages/com.unity.xr.arfoundation@5.0/manual/project-setup/install-arfoundation.html.

Now, if you revisit **XR Plug-in Management** via the aforementioned quick check, the **XR Simulation** option should be visible in the **Windows, Mac, Linux settings** tab, as shown in *Figure 4.15*.

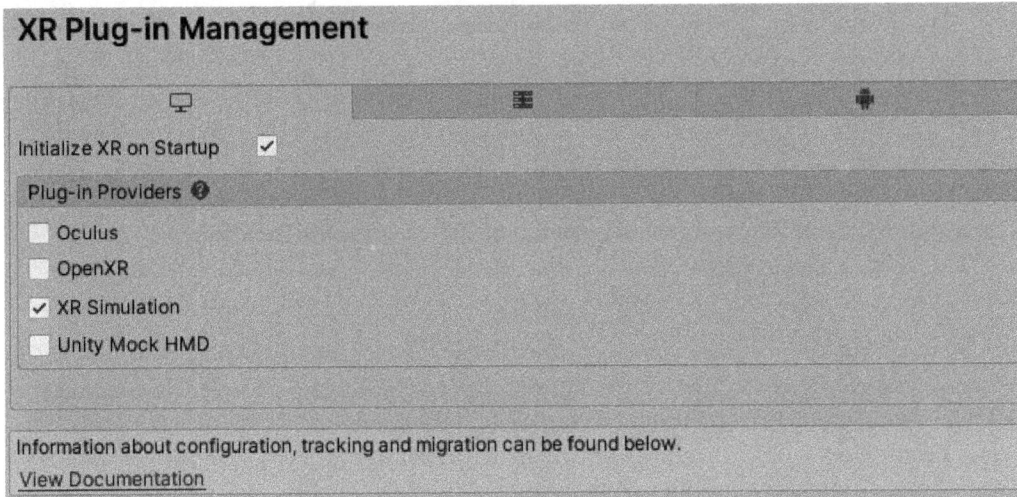

XR Plug-in Management

Initialize XR on Startup ✔

Plug-in Providers ❓

☐ Oculus
☐ OpenXR
☑ XR Simulation
☐ Unity Mock HMD

Information about configuration, tracking and migration can be found below.
View Documentation

Figure 4.15 – The Windows, Mac, Linux settings tab of the XR Plug-in
Management window with the XR Simulation checkbox enabled

With this feature activated, you're primed to choose an environment and test the AR scene within Unity.

Choosing an environment and testing the scene

To test an AR scene within Unity, you'll need an environment that emulates the real-world setting. To find the right simulation setting, go to **Window | XR | AR Foundation | XR Environment**. In the middle of the **XR Environment** interface, there's an **Environment** dropdown. Initially, your project will offer just one environment option. While it's possible to add more by importing sample environments, **DefaultSimulationEnvironment** is usually adequate for most testing needs. Simply choose this option. Once you've made your selection, hit the play button to activate play mode and start the simulation. You'll now be able to view your scene in real time, as depicted in *Figure 4.16*.

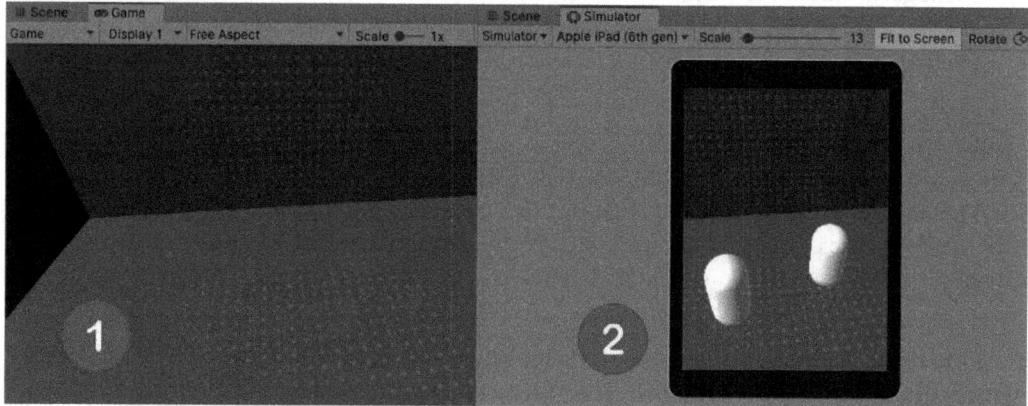

Figure 4.16 – XR Simulation at runtime

Pressing the play button transfers you to the **Game** mode of **XR Simulation**, presenting the simulated AR scene (highlighted as *1* in *Figure 4.16*). In this mode, you can adjust the viewpoint as needed, witnessing features such as plane detection in real time. Once satisfied with the viewpoint, you can shift from **Game** mode to **Simulator** mode (highlighted as *2* in *Figure 4.16*) using the dropdown in the top-left corner. It's important to switch to **Simulator** mode to access the **XR Simulation** functionalities, such as touch inputs. Observing the scene shown as *2* in *Figure 4.16*, you'll notice key features such as the distinct white point pattern, indicating the detected plane, and white capsules, marking the spots where mouse clicks occurred.

However, it's crucial to be aware of the constraints surrounding **XR Simulation**. Some elements might seem smaller than expected, and the resolution may not be the sharpest. Although **XR Simulation** provides a convenient testing avenue, it's not a complete substitute for deploying the AR scene on an actual device. Consider it a tool for iterative testing until you're confident about deploying it onto your target mobile device.

Now, let's proceed to launch the scene on our smartphone and observe the outcome.

Deploying AR experiences onto mobile devices

It's now time to launch your AR experiences onto smartphones or tablets. Primarily, you have two paths to accomplish this: deployment onto an Android or an iOS device.

For solo projects, where the AR application is meant for personal use, you may opt to deploy onto just Android or iOS, depending on your device's operating system. However, for larger-scale projects that involve several users – be it academic, industrial, or any other group, irrespective of its size – it's advisable to deploy and test the AR app on both Android and iOS platforms.

This strategy has multiple benefits. First, if your application gains momentum or its usage expands, it would already be compatible with both major platforms, eliminating the need for time-consuming porting later on. Second, making your app accessible on both platforms from the outset can draw in more users, and possibly attract increased funding or support.

Another key advantage to this is the cross-platform compatibility offered by Unity. This enables you to maintain a singular code base for both platforms, simplifying the management and updating process for your application. Any modifications made need to be done in one location and then deployed across both platforms.

In the next section, we'll delve into the steps required to deploy your AR scene onto an Android device.

Deploying onto Android

This section outlines the procedure to deploy your AR scene onto an Android device. The initial part of the process involves enabling some settings on your phone to prepare it for testing. Here's a step-by-step guide:

1. Confirm that your device is compatible with ARCore. ARCore is essential for AR Foundation to work correctly. You can find a list of supported devices at `https://developers.google.com/ar/devices`.

2. Install ARCore, which AR Foundation uses to enable AR capabilities on Android devices. ARCore can be downloaded from the Google Play Store at `https://play.google.com/store/apps/details?id=com.google.ar.core`.

3. Activate **Developer Options**. To do this, open **Settings** on your Android device, scroll down, and select **About Phone**. Find **Build number** and tap it seven times until a message appears stating **You are now a developer!**

4. Upon returning to the main **Settings** menu, you should now see an option called **Developer Options**. If it's not present, perform an online search to find out how to enable developer mode for your specific device. Though the method described in the previous step is the most common, the variety of Android devices available might require slightly different steps.

5. With **Developer Options** enabled, turn on **USB Debugging**. This will allow you to transfer your AR scene to your Android device via a USB cable. Navigate to **Settings | Developer options**, scroll down to **USB Debugging**, and switch it on. Acknowledge any pop-up prompts.

6. Depending on your Android version, you might need to allow the installation of apps from unknown sources:

    *   For Android versions 7 (Nougat) and earlier: Navigate to **Settings | Security Settings** and then check the box next to **Unknown Sources** to allow the installation of apps from sources other than the Google Play Store.

    *   For Android versions 8 (Oreo) and above: Select **Settings |Apps & Notifications | Special App Access | Install unknown apps** and activate **Unknown sources**. You will see a list of apps that you can grant permission to install from unknown sources. This is where you select the *File Manager* app, as you're using it to download the unknown app from Unity.

7. Link your Android device to your computer using a USB cable. You can typically use your device's charging cable for this. A prompt will appear on your Android device asking for permission to allow USB debugging from your computer. Confirm it.

With your Android device properly prepared for testing AR scenes, you can now proceed to deploy your Unity AR scene. This involves adjusting several parameters in the Unity Editor's **Build Settings** and **Player Settings**.

Here's a step-by-step guide on how to do this:

1. Select **File | Build Settings | Android**, then click the **Switch Platform** button. Now, your **Build Settings** should look something like what is illustrated in *Figure 4.17*.

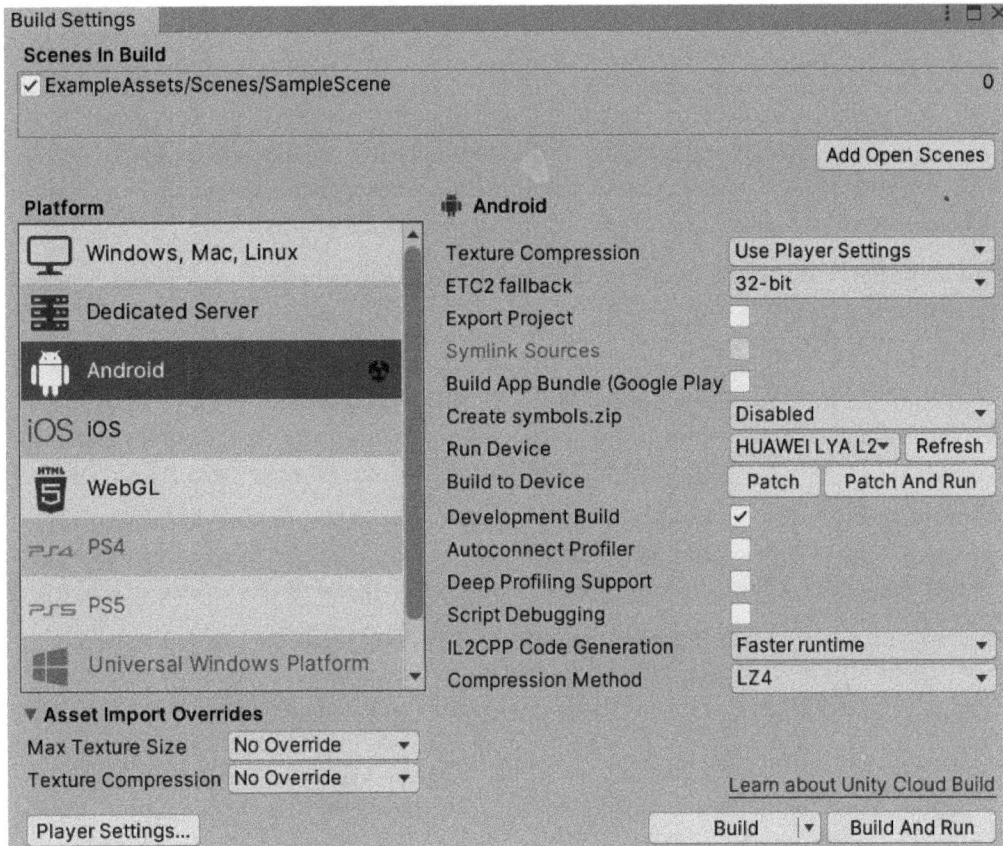

Figure 4.17 – Unity's Build Settings configuration for Android

2.   Next, click on the **Player Settings** button. This will open a new window, also called **Player Settings**. Here, select the **Android** tab, scroll down, and set **Minimum API Level** to Android 7.0 Nougat (API level 24) or above. This is crucial, as ARCore requires at least Android 7.0 to function properly.

3.   Remaining in the **Android** tab of **Player Settings**, enter a package name. Ensure it follows the pattern com.company_name.application_name. This pattern is a widely adopted convention for naming application packages in Android and is used to ensure unique identification for each application on the Google Play Store.

4.   Return to **File** | **Build Settings** and click the **Build and Run** button. A new window will pop up, prompting you to create a new folder in your project's directory. Name this folder Builds. Upon selecting this folder, Unity will construct the scene within the newly created Builds folder.

This is how you can set up your Android device for deploying AR scenes onto it. In the next section, you will learn how you can deploy your AR scene onto an iOS device, such as an iPhone or iPad.

Deploying onto iOS

Before we delve into the process of deploying an AR scene onto an iOS device, it's important to discuss certain hardware prerequisites. Regrettably, if you're using a Windows PC and an iOS device, it's not as straightforward as deploying an AR scene made in Unity. The reason for this is that Apple, in its characteristic style, requires the use of *Xcode*, its proprietary development environment, as an intermediary step. This is only available on Mac devices, not Windows or Linux.

If you don't possess a Mac, there are still ways to deploy your AR scene onto an iOS device. Here are a few alternatives:

- *Borrowing a Mac*: The simplest solution to gain access to Xcode and deploy your app onto an iOS device is to borrow a Mac from a friend or coworker. It's also worth checking whether local libraries, universities, or co-working spaces offer public access to Macs. For commercial or academic projects, it's highly recommended to invest in a Mac for testing your AR app on iOS.

- *Using a virtual machine*: Another no-cost alternative is to establish a macOS environment on your non-Apple PC. However, Apple neither endorses nor advises this method due to potential legal issues and stability concerns. Therefore, we won't elaborate further or recommend it.

- *Employing a Unity plugin*: Fortunately, a widely used Unity plugin enables deployment of an AR scene onto your iOS device with relatively less hassle. Navigate to **Windows | Asset Store**, click on **Search Online**, and **Unity Asset Store** will open in your default browser. Search for `iOS Project Builder for Windows` by Pierre-Marie Baty. Though this plugin costs $50, it is a much cheaper alternative than buying a Mac. After purchasing the plugin, import it into your AR scene and configure everything correctly by following the plugin's documentation (`https://www.pmbaty.com/iosbuildenv/documentation/unity.html`).

In this book, we focus exclusively on deploying AR applications onto iOS devices using a Mac for running Unity and Xcode. This is due to potential inconsistencies and maintenance concerns with other aforementioned methods.

Before you initiate the deployment setup, ensure that your Mac and iOS devices have the necessary software and settings. The following steps detail this preparatory process:

1. Make sure the latest software versions are installed on your MacOS and iOS devices. Check for updates by navigating to **Settings | General | Software Update** on each device and install any that are available.

2. Confirm that your iOS device supports ARKit, which is crucial for the correct functioning of AR Foundation. You can check compatibility at `https://developer.apple.com/documentation/arkit/`. Generally, any device running on iPadOS 11 or iOS 11 and later versions are compatible.

3. You will need an Apple ID for the following steps. If you don't have one, you can create it at `https://appleid.apple.com/account`.

4. Download the **Xcode** software onto your Mac from Apple's developer website at `https://developer.apple.com/xcode/`.

5. Enable **Developer Mode** on your iOS device by going to **Settings | Privacy & Security | Developer Mode**, activate **Developer Mode**, and then restart your device. If you don't find the **Developer Mode** option, connect your iOS device to a Mac using a cable. Open **Xcode**, then navigate to **Window | Devices and Simulator**. If your device isn't listed in the left pane, ensure you trust the computer on your device by acknowledging the prompt that appears after you connect your device to the Mac. Subsequently, you can enable **Developer Mode** on your iOS device.

Having set up your Mac and iOS devices correctly, let's now proceed with how to deploy your AR scene onto your iOS device. Each time you want to deploy your AR scene onto your iOS device, follow these steps:

1. Use a USB cable to connect your iOS device to your Mac.

2. Within your Unity project, navigate to **File | Build Settings** and select **iOS** from **Platform options**. Click the **Switch Platform** button.

3. Check the **Development Build** option in **Build Settings** for iOS. This enables you to deploy the app for testing purposes onto your iOS device. This step is crucial to avoid the annual subscription cost of an Apple Developer account.

> **Note**
>
> Deploying apps onto an iOS device with a free Apple Developer account has certain limitations. You can only deploy up to three apps onto your device at once, and they need to be redeployed every 7 days due to the expiration of the free provisioning profile. For industrial or academic purposes, we recommend subscribing to a paid Developer account after thorough testing using the **Development Build** function.

4. Remain in **File | Build Settings | iOS tab**, click on **Player Settings**, scroll down to **Bundle Identifier**, and input an identifier in the form of `com.company_name.application_name`.

5. Return to **File | Build Settings | iOS tab** and click **Build and Run**. In the pop-up window, create a new folder in your project directory named `Builds` and select it.

6. **Xcode** will open with the build, displaying an error message due to the need for a signing certificate. To create this, click on the error message, navigate to the **Signing and Capabilities** tab, and select the checkbox. In the **Team** drop-down menu, select **New Team**, and create a new team consisting solely of yourself. Now, select this newly-created team from the drop-down menu. Ensure that the information in the **Bundle Identifier** field matches your Unity Project found in **Edit | Project Settings | Player**.

7. While in **Xcode**, click on the **Any iOS Device** menu and select your specific iOS device as the output.

8. Click the **Play** button on the top left of **Xcode** and wait for a message indicating **Build succeeded**. Your AR application should now be on your iOS device. However, you won't be able to open it until you trust the developer (in this case, yourself). Navigate to **Settings | General | VPN & Device Management** on your iOS device, tap **Developer App certificate** under your **Apple ID**, and then tap **Trust (Your Apple ID)**.

9. On your iOS device's home screen, click the icon of your AR app. Grant the necessary permissions, such as camera access. Congratulations, you've successfully deployed your AR app onto your iOS device!

You now know how to deploy your AR experiences onto both Android and iOS devices.

Let's review what we have learned so far in this chapter before moving on to creating interactive XR experiences.

Summary

In this chapter, we've delved into the complexities and intricacies that surround AR glasses, exploring why these devices face numerous physical and technological challenges before they can be fully embraced by the public at large. We've examined the critical choice between marker-based and markerless approaches in your AR application, and we've discussed how this seemingly simple decision can significantly influence not only the development journey of your application but also its accessibility, versatility, and user engagement.

Through the exploration and installation of Unity's AR Foundation package, you are now empowered to create simple AR experiences of your own, and ready to deploy them across an extensive array of handheld, AR-compatible mobile devices.

We've also discovered that the deployment of an AR scene onto iOS devices can be a complex and time-intensive task, largely due to the various restrictions imposed by Apple's ecosystem compared to the Android ecosystem. However, these constraints should not deter you from pursuing a cross-platform approach for your AR apps. By aiming for deployment across both operating systems, you ensure greater accessibility and increase the potential for reaching a broader audience.

By understanding the diverse aspects of AR development through Unity, you are well on your way to creating immersive AR experiences that truly stand out. In the next chapter, you will learn how you can use C# scripting and other techniques to add complex logic to your VR applications.

5

Building Interactive VR Experiences

In this chapter, we delve into the world of more complex VR applications, empowering you to craft immersive and interactive experiences for a wide range of use cases. You will learn how to use the powerful Unity engine to create interactive VR apps without code and with C#.

The beauty of this chapter lies in its practical approach. You'll acquire hands-on skills to construct a VR environment that mirrors the responsiveness and realism of our natural world. Though some background in C# could be advantageous, it is by no means a prerequisite. Our tutorial is designed to be beginner-friendly, allowing anyone with an interest in XR applications to follow along comfortably.

This chapter is split into the following topics:

- Building interactive VR experiences without code
- Building interactive VR experience with C#

Technical requirements

Before we embark on this exciting journey, let's ensure you meet the needed technical requirements to power through this chapter. Besides having the most recent version of Unity installed, you also need to have either *Windows/Linux/Mac* or *Android Build Support* enabled, depending on whether your VR device is running standalone or in PC-based VR mode. If you are unsure whether your PC or laptop meets the hardware requirements of Unity, have a look at this page: `https://docs.unity3d.com/Manual/system-requirements.html`.

Building interactive VR experiences without code

After exploring the demo scene in *Chapter 3*, let's build interactive VR experiences ourselves, starting with a virtual car exhibition. In the experience we are going to build, the user should be able to walk around the ground via continuous movement (walking with the joystick) and via teleportation (only to teleportation anchors).

To kick off our experience, create a new project by following the *Setting up a VR project in Unity and the XR Interaction Toolkit* section in *Chapter 3*. If you have followed the steps in *Chapter 3*, you can just open the existing project for this chapter and go through the following steps.

Once your project is loaded, begin by creating a new empty scene. This can be done by navigating to **Assets | Scenes** in the project window, right-clicking, selecting **Create | Scene**, and naming it CarExhibition.

By double clicking on it, you're now inside the newly created scene that contains only a **Main Camera** and a **Directional Light**. You can delete the **Main Camera**, as it is not needed.

Before we delve into the VR Setup, we must first set the stage for our interaction. In this case, this means creating a dedicated area to house our two car exhibits. For the sake of simplicity, we'll be fashioning our exhibition ground from a simple cube primitive. Achieve this by right-clicking in the Scene Hierarchy and selecting **3D Object | Cube**. Position this cube at the origin and scale it to (20, 0.1, 20). Now, rename the cube ground. This will form the foundation of our car exhibition.

The next section details how we can import cars from the Asset Store.

Importing cars from the Asset Store

Animating cars in our scene requires the prerequisite step of importing them. Let's walk through the process of importing the simple cars pack from the Unity Asset Store:

1. Go to the Unity Asset Store (https://assetstore.unity.com/). Search for the simple cars pack or, for a more direct route, access the package page via this link: https://assetstore.unity.com/packages/3d/vehicles/land/simple-cars-pack-97669.

2. After adding the package to your assets, find the **Open in Unity** button and click on it.

3. This action should prompt the Unity application to open, presenting you with the option to import the package. Please proceed with the **Import** operation.

4. Once the import is successful, journey to the package's directory in the project window. The path you're looking for is **Assets | Simple Vehicle Pack | Prefabs**.

5. Unleash the taxi into the scene at coordinates (-6,0,0) and adjust its scale to (2,2,2).

Occasionally, you might encounter a peculiar magenta hue on the object. This typically implies that the associated materials are missing or incompatible with the current rendering pipeline. To rectify this issue, go through the following steps:

1. *URP Installation*: Before proceeding, ensure you have the URP installed, as it's not included by default in the VR Template project. If it's not installed, go to **Window | Package Manager**. Then, search for the URP and install it.

2. Head to `Assets | Simple Vehicle Pack | Materials | Cars_1`.

3. With the `Cars_1` material selected, alter the **Shader** drop-down menu in the **Inspector** window to **URP | Lit**.

4. Then, click on the small torus symbol next to **Base, Metallic,** and **Occlusion Map**, assigning the corresponding `cars_albedo` to the **Base Map**, `cars_metallic` to the **Metallic Map**, and **cars_AO** to the **Occlusion Map**.

   This should restore the car's normal appearance.

5. Repeat this process for `Bus_1`, `Bus_2`, and `Cars_2` materials, ensuring the corresponding `bus_albedo`, `bus_albedo_2`, `bus_metallic`, `bus _AO`, and `cars_albedo_2` textures are assigned to the relevant maps.

Note

You may have noticed that making manual adjustments can be quite time-consuming. Fortunately, Unity offers a more streamlined solution for such scenarios. The **Render Pipeline Converter** is designed to handle bulk conversions of assets and shaders to the URP format. Here's how to use it:

1. Go to **Window | Rendering | Render Pipeline Converter**.

2. In the opened window, select the materials or shaders you want to convert.

3. Click **Convert**. The tool will then attempt to automatically adjust your selected materials or shaders for compatibility with URP.

Do keep in mind that while this is convenient for bulk operations, you might still need to verify each material to ensure it is converted correctly.

With this task completed, revisit the `Assets | Simple Vehicle Pack | Prefabs` and let your creativity run wild by placing all the vehicles from the asset into our scene. Please note that in addition to the Taxi, we're particularly interested in the `Police_car`, positioned at (6,0,0).

Now that our cars have arrived at the scene, we're ready to add the player and button prefabs in the next section.

Adding the player, teleport anchors, and button to the scene

Before we begin implementing our game logic into the VR experience, we must first add some additional GameObjects and prefabs to our scenes that are just as important for its functionality as the cars themselves. Let's start with adding the player.

Adding the player

Let's imagine our scene as a stage and ourselves as the directors. In our toolbox, we've got this lovely XR Interaction setup prefab, a sort of puppet we can animate around our virtual stage. We highly recommend using the XR Interaction Toolkit version *2.5.1* by following the *Installing the XR Interaction Toolkit and Samples* section in *Chapter 3*. While the toolkit is consistently updated, newer versions might come with different prefabs. Let's add and configure this prefab using the following steps:

1. The XR Interaction setup prefab is easy to find; it's nestled in the `Assets | Samples | XR Interaction Toolkit` folder. Search for the version you have, maybe it's *2.5.1*, as recommended, or something newer. Then, dig through **Starter Assets** till you find **Prefabs**.

2. Get the **XR Interaction Setup** in the `Prefabs` folder and drag it right on the scene. Adjust it as needed to sit right at the center, at coordinates (0,0,0).

Successfully, we've integrated a player into our scene. Now, let's enhance the experience by setting up teleport anchors at various points of interest within our environment.

Setting up two teleport anchors

For those new to VR, a great way to prevent motion sickness is to restrict movement to teleportation. The XR Interaction Toolkit provides us with three teleportation prefab options: **Teleport Anchor**, **Snapping Teleport Anchor**, and **Teleportation Area**. Think of **Snapping Teleport Anchor** as a specific point on our stage where the teleportation ray snaps to. In contrast, with **Teleport Anchor**, the teleportation ray will glide without snapping. However, both anchors essentially serve the same purpose: they're designated spots where users can teleport. On the other hand, **Teleportation Areas** are expansive zones allowing user teleportation within their boundaries.

For this particular scene, we will be using two snapping teleport anchors to ensure users teleport only to specific locations. Here's how we can set them up:

1. In our project window, navigate to `Assets | Samples | XR Interaction Toolkit | 2.5.1 | Starter Assets | DemoSceneAssets | Prefabs | Teleport`. Within this location, you'll come across the just-described **Teleport Anchor**, **Snapping Teleport Anchor**, and **Teleportation Area**.

2. Drag the snapping teleport anchor from the `Teleport` folder and drop it into the **Scene Hierarchy** window. Rename this to `Taxi Teleport Anchor`. Position it at coordinates (-2, 0, 0) and scale it to (2, 1, 2).

3. Drag and drop another snapping teleport anchor into the **Scene Hierarchy** window. Rename this one to `Police_car Teleport Anchor`. Position it at (2, 0, 0) and scale it to the same size as the previous anchor: (2, 1, 2).

Once set up, your scene should resemble what's illustrated in *Figure 5.1*.

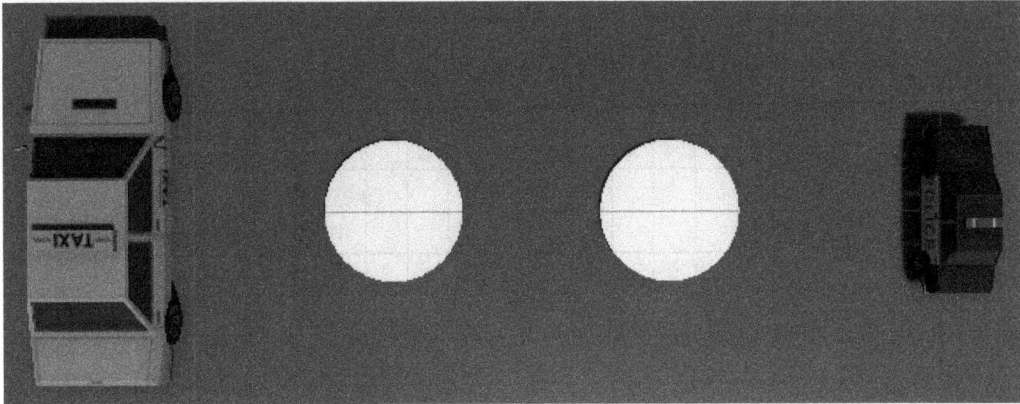

Figure 5.1 – The taxi and the police car with their teleport anchors

Next, we'll enhance the `Taxi Teleport Anchor` by adding a push button directly onto it.

Adding a push button for the taxi animation

To make our push button for the taxi animation stand out and be easily accessible, we're going to place it on a pedestal created using a cylinder. This cylinder will serve as the button's dedicated stand:

1. To create this pedestal, right-click in the **Scene Hierarchy** window and select **3D Object** | **Cylinder**. Rename this cylinder `Taxi Button Stand`. Adjust its size to dimensions (0.2, 0.5, 0.2) and position it at (-2.7, 0.55, 0).

2. To find the **push** button, return to `Assets | Samples | XR Interaction Toolkit | 2.5.1 | Starter Assets | DemoSceneAssets | Prefabs | Interactables`. Once you've found it, drag and drop it onto the `Taxi Button Stand` in the **Scene Hierarchy** window. Make sure to scale it (5, 5, 5) and adjust its y-coordinate to 1.075 so it sits correctly on the pedestal.

Refer to *Figure 5.2* to ensure that your scene looks as expected with the newly added push button. In the subsequent section, we'll focus on setting up text buttons for the police car animations.

Figure 5.2 – The taxi button stand we just added to the scene

Adding text buttons for police car animations

For our police car, we'll create a canvas that holds two buttons, enabling users to either shrink or enlarge the police car.

Setting up the canvas

1. Begin by creating a canvas: right-click in the hierarchy and select **XR | UI Canvas**.
2. Adjust its size dimensions. The correct size for our purpose is (0.001, 0.001, 0), making it a square. This scaling is essential for optimal viewing in the VR environment.
3. Position the canvas at coordinates (3,1,-1) and rotate it (0, 90, 0) towards the player's direction.
4. If you're facing difficulties altering the UI Canvas properties, head to the Inspector window of the canvas component. Here, change the **Render Mode** from **Screen Space – Overlay** to **World Space**.

Adjusting the Canvas components

1. With the canvas selected in the **Scene Hierarchy**, open the **Inspector** window. Click on **Add Component** to attach new components to the canvas.
2. First, add a **Vertical Layout Group** component. This ensures that any buttons you subsequently add will automatically align on top of each other. Within the **Vertical Layout Group** settings, do the following:

    I.   Set **Spacing** to 5.

    II.   Choose **Middle Center** for **Child Alignment**.

III.   Check **Child Force Expand** for both width and height.

IV.   For visual appeal, add an **Image** component to the canvas. In the **Image** component's settings, set the source image to **UISprite** and choose a background color that complements your scene.

Adding the buttons

1.   Navigate to `Assets | Samples | XR Interaction Toolkit | 2.5.1 | Starter Assets | Prefabs | DemoSceneAssets | Prefabs | UI`. Here, you should find the **TextButton** prefab.

2.   Drag and drop this prefab onto the **Canvas** twice. This will automatically place two **TextButton** prefabs as child objects beneath the **Canvas**. Due to the **Vertical Layout Group**, they'll be aligned neatly.

3.   Scale them to (`0.4, 0.4, 0`) and rename the first button as `Scale Big Button` and the second one as `Scale Small Button`.

Labeling the buttons

Modify the child text object of each button by unfolding them twice. With the `Text GameObject` selected, head to the Inspector window and search for the text component. Label one button as `Scale Big` and the other as `Scale Small`.

By following these steps, you'll have a neatly organized UI ready for your police car animations, as shown in *Figure 5.3*.

Figure 5.3 – The text buttons for the police car animations

Now our stage is set, it's ready for the show and should look like *Figure 5.4*.

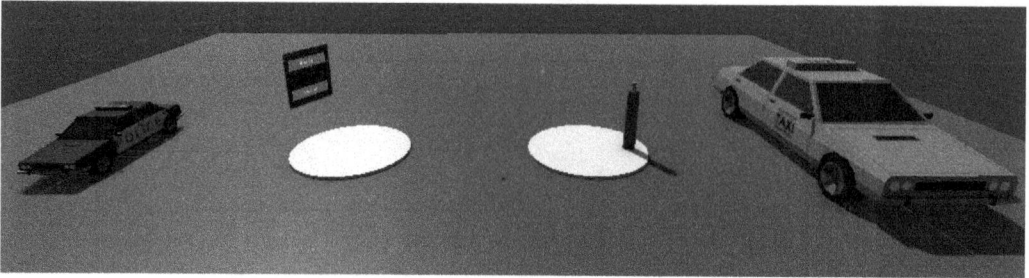

Figure 5.4 – The current status of the CarExhibition scene

Up next, we're going to learn about the puppet strings, or in other words, the button events we can manipulate in Unity.

Interactable events that can be triggered

Interactable events, particularly those found in the **XR Simple Interactable** component, play a pivotal role in enhancing user experience. Such events, often tied to push buttons, facilitate diverse interactions within VR environments. There are many different options for interactable events. Imagine you are in a virtual reality car exhibition where you can interact with push buttons next to each car. Here's how these events might apply in this scenario:

- **First Hover Entered**: You approach a car and point at its information button for the first time. This event triggers a subtle glow around the button indicating it is selected.

- **Hover Entered**: Each time you point at the button, the event renews the glow, showing that the button is currently the focus of your attention.

- **Hover Exited**: As soon as your hand or pointer moves away from the button, the glow dissipates, indicating that the button is no longer selected.

- **Last Hover Exited**: The final time you move away from the button, the glow dissipates, and it seems as though the button subtly moves back to its original position, indicating that it's no longer in focus.

- **First Select Entered**: The first time you press the information button, this event is triggered. It might cause the car's information panel to appear with a smooth animation.

- **Select Entered**: Each time you press the button, the car's information panel appears, whether it's the first press or not.

- **Select Exited**: When you release the button, the panel begins to fade, indicating the end of the interaction.

- **Last Select Exited**: The final time you release the button, it seems as though the panel not only fades but also shrinks back into the button, signifying the interaction's conclusion.

- **Activated**: This event could be tied to a button in the car that starts the engine. Upon pressing, you hear the car's engine roar to life.

- **Deactivated**: When you press the button to turn off the car's engine, this event is triggered. You hear the engine wind down to a stop.

This sequence of events provides a natural, intuitive interaction flow for the user, enhancing the immersive feel of the VR car exhibition.

In the next section, we'll delve into how to choreograph these interactions and bring animations to life without writing a single line of code.

Understanding animations and animator systems

Unity's **animation system** functions like a mechanism for manipulating GameObjects by controlling their movement, rotation, size, color, and so on. It works by defining keyframes, which are specific states at certain times, and constructing an animation clip from these keyframes.

Consider the example of a virtual door. An animation clip can illustrate the door's movement from closed to open, and another clip can depict the door returning to its closed position.

To ensure smooth transitions between these clips, Unity's **Animator Controller** is used. This is a control system that directs the timing and sequence of the animation clips, transitioning between them based on specific conditions.

For the door example, the Animator Controller would manage the clips for opening and closing the door and transitioning between them at the right moment, possibly based on user input.

Here are other examples that can be crafted using Unity's animation system:

- **Object movement**: For an object moving across the screen, the animation system can create the movement, but additional scripting may be required to control it based on user interaction

- **Object rotation**: A constantly spinning gear can be animated, but player-based control of the spin requires code

- **Object scaling**: Scaling animation can depict an object's growth, but linking this to other factors, such as energy level, requires scripting

- **Color change**: An animation can be made for a changing light color, but synchronizing it with game conditions requires coding

- **Camera movement**: Camera movement between characters can be created using keyframes, but complex interactions with player movement might need additional code

- **Blend shapes**: Facial expressions can be controlled using blend shapes, but they may need to be scripted to correspond to specific game dialogues

- **UI animations**: A graphical UI element, such as a flashing button, can be animated, but stopping the animation when certain conditions are met (e.g., a race begins) requires a script

As seen in these examples, the animation system in Unity provides a powerful tool for bringing objects and scenes to life. However, adding scripts or coding to these animations unlocks a higher level of refinement and interactivity.

In general, coding enables us to create animations in Unity that can respond to various inputs and game states, making them more dynamic. Coding also provides precise control over the animation, allowing for complex logic and customized interactions that cannot be achieved with Unity's animation system alone.

While Unity's animation system may not have the same level of complexity and fine control that coding offers, it still holds a significant place as a primary tool for creating interactions within Unity. The visual interface and intuitive controls allow for a rapid and user-friendly way to animate GameObjects, even without delving into scripts. This makes it accessible to both novice and experienced developers, enabling them to add life and movement to their game scenes.

In the next section of this chapter, we will only be using the animation system, demonstrating how powerful and useful it can be on its own. Through hands-on examples, you'll see how it's possible to achieve engaging and interactive animations without relying on coding and how this tool can become a vital part of your XR development toolkit.

Animating a 360-degree car rotation

Let's dive into the world of animation. Think of a spinning top: it goes round and round, and we want our taxi to do the same when we push a button. The twist is that when we let go of the button, the taxi should stop spinning. Sound fun? Here's how we'll make it happen using the magic of Unity's animation system and UnityEvents without writing a line of code.

First things first, we need to choreograph this spin for our taxi. Imagine it as a pirouette, a 360-degree turn on the taxi's y-axis. Go through the following steps to bring this dance to life:

1. Right-click in the `Assets` folder and select **Create | Folder**.

2. Select the taxi object in your scene and go to **Window | Animation | Animation** to bring up the Animation window. Click on **Create**, as if you're penning a new script for our taxi's dance, name it `RotateCar`, and save it in our newly minted `Animations` folder.

3. Define a 360-degree rotation for the y-axis by selecting **Transform | Rotation** and modifying your keyframes as shown in *Figure 5.5*.

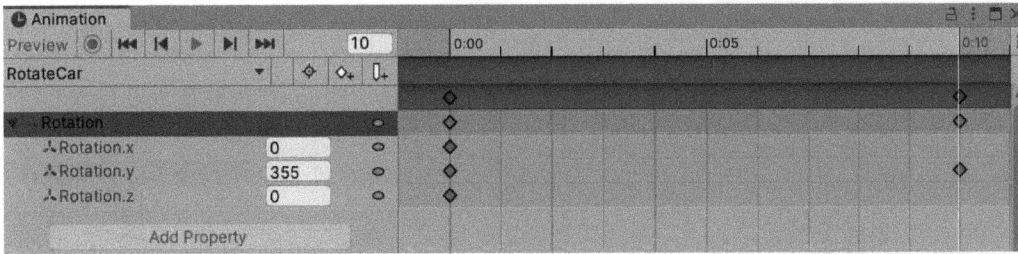

Figure 5.5 – The modified Keyframes to define a 360-degree rotation for the y-axis

Set the initial keyframes at 0:00 with 0 as the rotation; this is our taxi's starting pose. Then, leap to 0:10 and give the **Rotation.y** value a 355-degree spin. Why not 360? That would be like turning on a dime, too abrupt for our gentle spin.

4. Click on the RotateCar animation clip in the project window to view its settings in the Inspector. In the Inspector, give the **Loop Time** box a check to tell our taxi to keep repeating its spin.

5. Attach an **Animator Component** to the Taxi and drag the newly created RotateCar animation clip into the **Controller** field of the **Animator** component.

This will start our Taxi's twirl right when we step into our virtual world. But remember, we want the spin to start only when we push the button. It's time to make that happen:

1. First, we need to open the Animator Controller. Select the **Animator Controller** for our Taxi and double-click it to open the Animator Controller window.

2. Like a blank canvas, right-click in the **Animator** window and select **Create State | Empty** to create a new empty state. Name this state Idle; it's the calm before the Taxi's spinning storm.

3. Next, right-click on the Idle state and set it as the **Layer Default State**. This tells the Animator Controller to start in this state, doing nothing when the scene begins.

4. We then make a bridge from the Idle state to the RotateCar state: the state that initiates our taxi's spin. To do this, right-click on Idle, choose **Make Transition**, and then click on RotateCar.

5. Now let's navigate to the **Parameters** tab and add a parameter named Rotate. This parameter will initiate the transition from the Idle state to the RotateCar state.

6. Click on the arrow representing the transition from Idle to RotateCar. In the **Inspector** window, under **Conditions**, add a new condition. Choose the Rotate parameter that we just created, as shown in *Figure 5.6*.

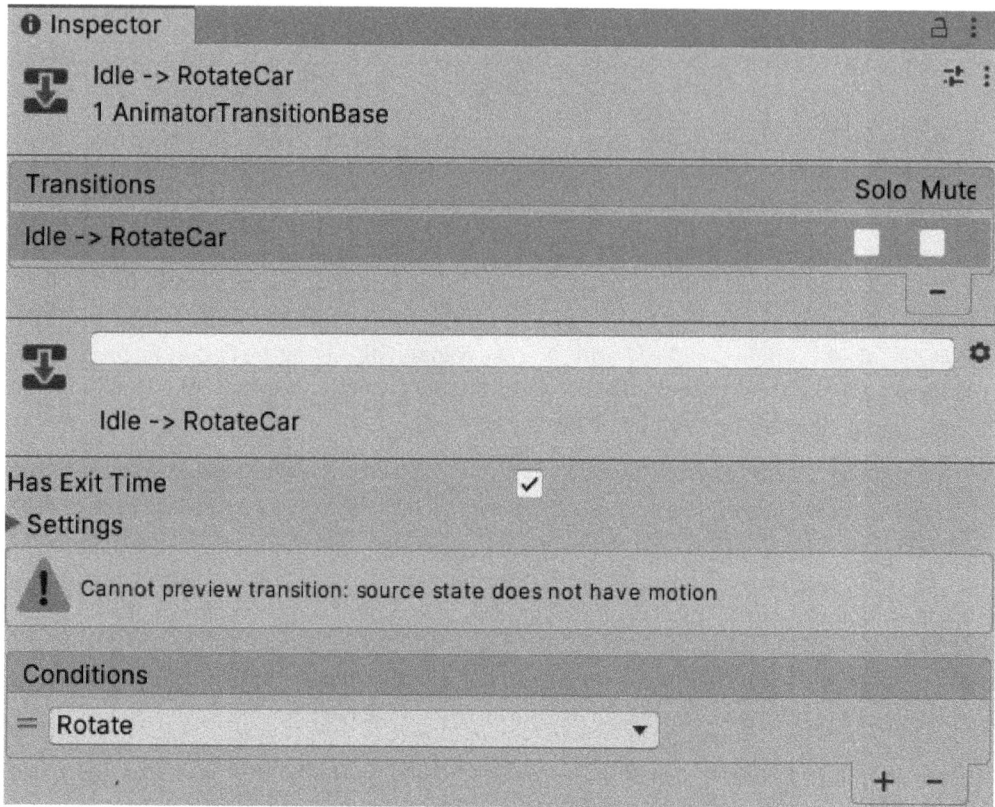

Figure 5.6 – The Inspector window showing the newly added condition: the Rotate trigger

7. Next, we need to halt our taxi's spin once the button is released. Right-click on the RotateCar state, choose **Make Transition**, and then click on the Idle state.

8. Go back to the **Parameters** tab and this time add a parameter named StopRotation. This will initiate the transition from the RotateCar state back to Idle. Your parameters should now match what's shown in *Figure 5.7*.

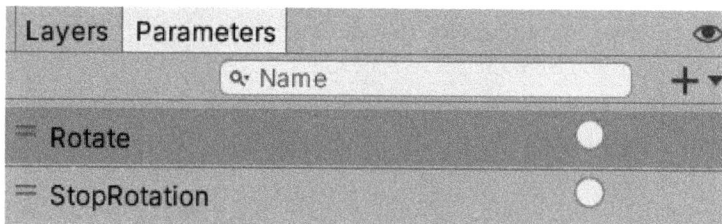

Figure 5.7 – The Parameters tab of the Animator window showing both newly added parameters

9. Click on the transition arrow that goes from the `RotateCar` to the `Idle` state this time. Under **Conditions** in the Inspector, add a new condition and select the `StopRotation` parameter as we did with the `Rotate` parameter earlier. Your **Animator** window should now resemble *Figure 5.8*.

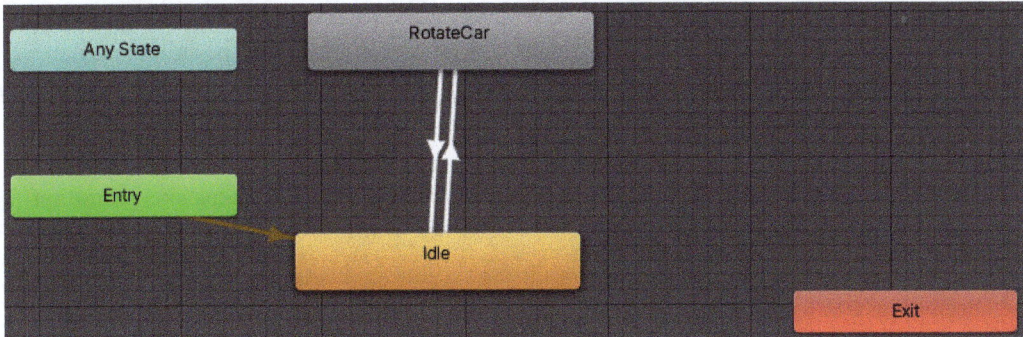

Figure 5.8 – The Animator window showing the current states and their transitions

10. Select the **Taxi push button** in the hierarchy and look for the **Select Entered** and **Select Exited** events. Click on the + and drag the taxi into the **Runtime Only** field. Next, we are calling for `Animator.SetTrigger()`. Think of it as the cue for our Taxi to start or stop its dance. When the button is pressed (**Select Entered** event), the cue is `Rotate`, and when the button is released (**Select Exited** event), the cue is `StopRotation`. If set up correctly, your events will mirror those in *Figure 5.9*.

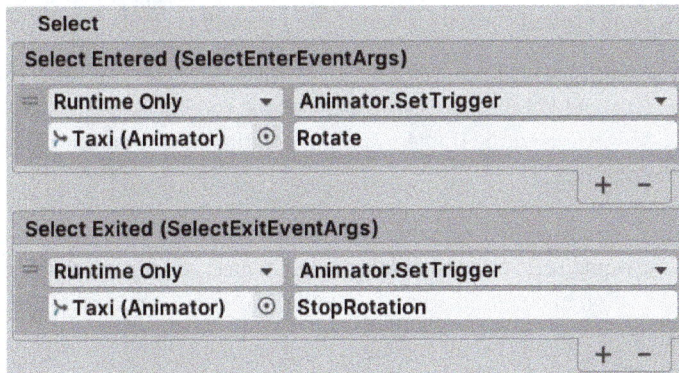

Figure 5.9 – The Select Entered and Select Excited events

Well done! Your taxi is now set to dance to your button's beats, spinning a full circle each time you press it. Don't forget to try it out: give that button a good press and watch the taxi perform its 360-degree spin.

In the next section, we'll delve into the magic of 2D text buttons on a canvas, learning how they can change a police car's size to make it shrink or grow at the press of a button.

Scaling a police car

Imagine the compelling possibilities of incorporating two additional interactive features in our VR auto show: one button to miniaturize the car and another to magnify it. Similar to our earlier interactions, we will employ a set of procedures, but this time with 2D buttons on a canvas. First, we create two animation clips that capture the essence of the car's transformation as it scales up and down. Next, we create an Animator Controller: a crucial component that defines and manages the animation states. Lastly, we assign the corresponding trigger events to the button.

Let's bring this magic to life with the following steps:

1. Select the police car in the hierarchy. Open the Animation window (**Window** | **Animation** | **Animation**) and create a new animation clip. Name it `ScaleCar` and store it in the `Animations` folder.

2. Now, let's choreograph the transformation. On the timeline in the Animation window, mark the beginning (`0:00`) and the end (`1:00`) with keyframes (the diamond icon). At the beginning, the scale of the car (**Transform** | **Scale**) should be at its smallest size, `0`. At the end, the scale should be at its largest size; let's go with `3`. This gives us an animation that sees the car scaling from minuscule to massive.

3. We don't want our car continuously growing and shrinking. So, in the Inspector, uncheck the **Loop Time** box.

4. Our police car needs an **Animator Component** to manage its transformation. Attach one if it hasn't been done automatically. Then, slot the `ScaleCar` animation clip into the **Controller** field of the **Animator Component**.

5. We've got the growing part down, now let's work on the shrinking. Create a second animation clip named `ShrinkCar`.

6. Similar to *step 2*, mark the beginning and the end with keyframes. This time, at the beginning, the scale of the car should be at its largest size (`3`). At the end, the scale should be at its smallest size (`0`). We've now reversed the transformation, creating an animation that shrinks the car from massive to minuscule.

By now, we've successfully created two animation clips and an Animator Controller. If we were to play the scene, the police car would scale up once thanks to the unchecked loop checkbox and the absence of the downscaling animation in our Animation Controller. We're one step closer to our transforming car. Ready for the next part? Let's dive in.

Now, it's time to make our police car dance to our tune. We want it to change sizes when we wish and be idle when we don't. That's where the Animator Controller comes in. It's our maestro, dictating when the car grows, shrinks, or takes a breather. Here's how we bring this to life:

1. First, we need to open the Animator Controller. Locate the one assigned to the police car in the project window and double-click to open it.

2. Remember our shrinking dance? Drag and drop the `ShrinkCar` animation clip into the **Animator Controller** window. It now has both the growing and shrinking routines.

3. Now, we need a state of rest, a breather in between our dance routines. Right-click in the **Animator Controller** window, select **Create State | Empty** and name it Idle. It's like the calm before the storm.

4. We don't want our car to start dancing right away when the curtain rises. So, set the Idle state as the default state. This ensures that when the scene begins, our car stands still, awaiting its cue.

5. Our maestro is ready to conduct. Right-click on the Idle state and choose **Make Transition**. Then, click on the ScaleCar and another time to the ShrinkCar states. Now, we've set the stage for our car to transition from standing idle to growing big and shrinking small.

6. But what cues the transitions? We need to set trigger parameters for this. In the **Animator** window, click on the **Parameters** tab and add two new trigger parameters:

    I.   Name them ScaleBig and ScaleSmall.

    II.   Select the transition arrows. Under **Conditions** in the **Inspector** window, add a new condition for each of these triggers.

    ScaleBig cues the transition from Idle to ScaleCar, while ScaleSmall cues the transition from Idle to ShrinkCar.

    The dance is never one-way. Our car needs to move smoothly from growing to shrinking and vice versa. So, create transitions from ShrinkCar to ScaleCar and the other way around. Click on the transition arrows. Under **Conditions**, add the ScaleBig trigger for the lower and the ScaleSmall trigger for the upper transition arrow, as shown in *Figure 5.10*.

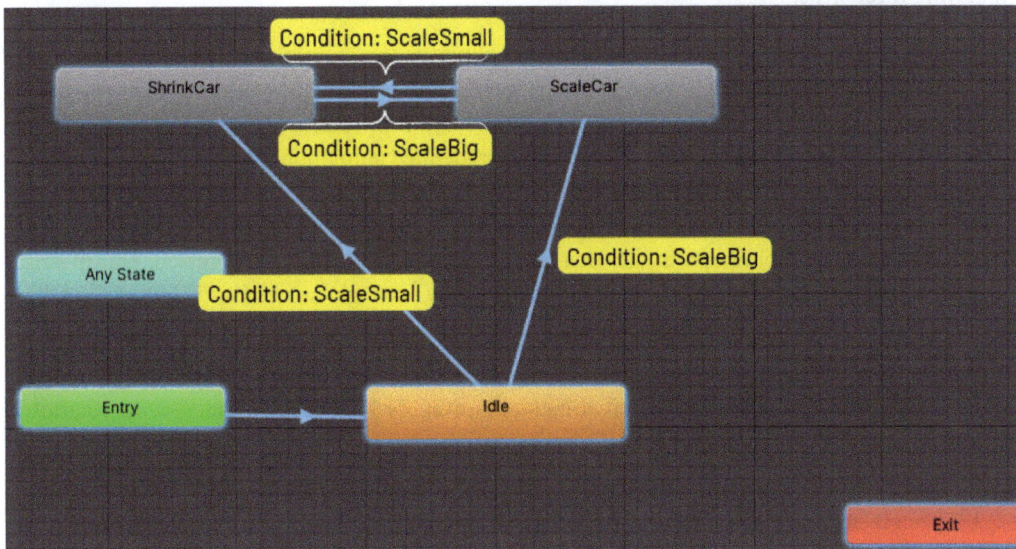

Figure 5.10 – The Animator Controller of the police car

Voila! Our police car is now ready to twirl to our commands. Try out the scene and watch as it gracefully grows and shrinks at your will by testing or deploying the scene onto your VR headset, as described in the *Deploying and testing VR experiences onto different VR platforms or simulators* section in *Chapter 3*. It's a delightful sight, isn't it?

Building interactive VR experiences with C#

So far, we have added interactions to our scene just by using Unity's Animation system without writing a single line of code. In the next sections, you will learn how you can use scripting with C# to add even more complex breaths of air into your GameObjects and scene.

The first question we are going to answer is this: when do we need to write C# code for our animations and interactions?

Understanding when to use C# for animations and interactions

It's important to understand that Unity's animation and animator systems and the use of C# scripting are not mutually exclusive. They are often used together, with the animator controlling predefined animations and C# adding interactivity based on user input or other game events.

In the previous section, we rotated and scaled cars based using Unity's animation and animator systems. These systems are primarily used to create predefined animations. We can divide this process into the following three steps:

1. **Initialization event**: First, an event must occur to start the animation. This can be a button press, game event, collision, or any other trigger. In our example, to scale and rotate the cars, we used a physical push button and a 2D UI button. Unity's built-in animation and animator systems are largely designed around the idea of predefined animations that are triggered under specific conditions. These conditions are typically defined within the Animator Controller itself and are usually based on parameters that you set up ahead of time, such as a boolean to track if a character is jumping or a trigger that gets activated when a button is pressed. This means that, with the built-in system, we do not have the ability to trigger animations based on virtually any event or condition in our game and have to work with the available events.

2. **Animation**: Using Unity's animation system, we define what changes during the animation (position, rotation, scale, etc.) by creating keyframes. For instance, we changed the rotation and scale of the cars with keyframes.

3. **Animator Controller**: This is where you manage the different animations and transitions between them. You create states, each linked to a different animation, and define the conditions under which transitions occur. For example, we used a button press that initialized a transition of the car from a `ShrinkCar` state to a `ScaleCar` state.

This system is perfect for creating animations that are predefined and occur under specific conditions; for instance, character animations (walking, running, jumping) or environmental animations (door opening, elevator moving).

While Unity's built-in system excels at predefined animations, C# comes into play when animations need to respond dynamically to user input or other game events. Here, we follow similar steps as before:

1. **Initialization event:** The event that starts an animation can be anything, such as button presses, user input, or changes in the game state. The biggest difference with Unity's built-in system is that with C# scripts, you have much greater flexibility and control over the initialization events for your animations. For instance, you can respond to complex sequences of input, such as a fighting game action combo. You can base animations on game logic or game state, such as an enemy's health level or the player's current score. You can trigger animations based on collisions, entering/exiting certain zones, or other physics events.

2. **AI Decisions:** You can initiate animations based on the decisions made by an AI system. In a multiplayer game, you can trigger animations based on network events, such as another player's actions.

3. **Animation:** Animations can be created in two ways:

   - You can use the animation system to create keyframes as before and then use C# to control the playback of these animations (for example, play, pause, stop, or alter speed).

   - Alternatively, you can use C# to modify the properties of GameObjects directly, creating animations programmatically.

4. **State control:** With C#, you gain more direct control over when and how animations and transitions occur. You can create conditions based on any aspect of your game's state, not just parameters in the Animator Controller. For example, you could change an NPC's animation based on the player's health or inventory.

Using C# for animations is ideal when the animation needs to respond in complex ways to the game state or user input. This is especially the case in the following scenarios:

- **Real-time user input:** You often use C# to animate GameObjects in response to real-time user input. For example, in a flight simulator, you might animate the plane's control surfaces (such as the ailerons, elevators, and rudder) based on the player's joystick input. Since the exact position of these surfaces depends on the player's input, it's not something that can be predetermined with keyframes.

- **Physics-based animations:** If your animation needs to incorporate or respond to physics, it often makes sense to animate it programmatically. For example, in a pool game, the balls move and spin based on physics calculations rather than predefined paths.

- **Procedurally generated content**: When the content of your game is procedurally generated, you often need to animate things programmatically because the exact nature of the animations can't be predetermined. For instance, in a rogue-like dungeon crawler, the layout of the dungeon and the placement of enemies are generated on the fly, so any animations related to these elements would also need to be generated at runtime.

- **Complex AI behavior**: When creating complex AI behavior, you might use C# to animate GameObjects based on the AI's decision-making processes. For instance, an enemy character might have an idle animation, a walk animation, and an attack animation, and you can use C# to decide which one to play based on the AI's current state and the player's position.

Now that we have understood the scenarios in which it is useful to use C# for animations and interactions, let's put our new knowledge to use by scaling a car using a slider in the following section. This would fall under the "real-time user input" category. When we want to scale a car using a slider, it's the user input (moving the slider) that's driving the animation (the scaling of the car). This kind of interaction can't be predefined with keyframes because the exact scale of the car depends on the player's input at any given moment. But before we add these to our scene, we first need to understand the very basics of the C# language, which are explained in the following section.

Understanding scripting with C# in Unity

Creating scripts for VR development in Unity involves using C#. As a high-level language, it simplifies many computing complexities, making it user-friendly compared to languages such as C++.

Through its strong, static typing, C# helps you spot programming errors before running a game in Unity, which is incredibly helpful for developers. Additionally, it's good at managing memory use, as it minimizes the risk of **memory leaks**—a situation where a game eats up an increasing amount of memory, potentially causing crashes.

C# is supported by Microsoft, which ensures you get reliable help, plenty of tools to work with, and access to a big community of other developers. And, importantly, just like Unity, C# works across many platforms.

When using C# in Unity, your coding is mostly event-based. This means you override certain Unity functions that get triggered at specific times, such as the `Start()` or `Update()` functions of a game.

Now, when it comes to using C# for VR development in Unity, there are three key object-oriented programming concepts you'll need to understand: variables, functions, and classes. Let's get to know these a bit better.

Variables

Variables are like storage boxes that your script uses to hold data. Every variable has a type, which tells you what kind of data it can hold. For example, if you're using the XR Interaction Toolkit, you might have variables to store things such as the position of a VR controller, the state of an object in the virtual world, and so on. Let's see an example of how we can use variables in the context of a C# script for Unity with the following code snippet:

```
public class XRGrab : MonoBehaviour
{
    public XRGrabInteractable grabInteractable;
    private bool isGrabbed = false;
}
```

In the preceding code, `grabInteractable` is an `XRGrabInteractable` object that represents a VR object that can be interacted with. `isGrabbed` is a private Boolean variable tracking whether the object is currently grabbed or not.

Functions

Think of **functions** (or methods) in C# as cooking recipes. Just like a recipe provides step-by-step instructions to cook a specific dish, a function in C# consists of a set of instructions that performs a specific task. You can reuse these recipes multiple times, either within the same script or across different scripts.

In Unity, there are some special functions, such as unique cooking recipes, that are triggered at specific times during the life cycle of a script. Here is an overview of them:

- `Awake()`: This is like an alarm clock for your script. This function rings when the script first wakes up (or loads). Often, it's used to set up variables or the state of the game.

- `Start()`: This is the runner on their mark, ready to start the race. This function gets called right before the first frame of the game is displayed. It's also used to set things up, but unlike `Awake()`, it won't run if the script isn't enabled.

- `Update()`: This function is the heart of your game, beating once per frame. It's usually used for tasks that need to happen regularly, such as moving objects around, checking for user input, and so on.

- `FixedUpdate()`: This is like `Update()`, but it runs at a consistent pace, no matter the frame rate. It's typically used for physics-related tasks.

- `LateUpdate()`: This is the function that tidies up after everyone else. It runs after all `Update()` functions have done their thing, doing any tasks that need to happen after everything else.

- `OnEnable()`: This method is a special Unity method that gets called whenever the object this script is attached to becomes active in the game.

Now, when it comes to VR interactions, Unity's XR Interaction Toolkit provides its own set of special functions that react to VR actions. Here are a few important ones you'll want to know about:

- `OnSelectEntered()`: This function is called when you select an interactable object in VR. Imagine it as the moment when you point at a virtual object with your VR controller.

- `OnSelectExited()`: This function is the moment when you stop selecting an interactable object. Think of it as letting go of the object you were pointing at.

- `OnActivate()`: This is when you activate an interactable object. It's a bit like pressing a button while you're already holding the object.

- `OnDeactivate()`: This is when you deactivate the object. It's like letting go of the button you just pressed.

- `OnHoverEntered()`: This function gets called when you start hovering over an interactable object. This could be used to make the object light up, for instance.

- `OnHoverExited()`: This function is when you stop hovering over an object.

These functions can be combined in different ways to create all sorts of interactions in the virtual world. It's like stacking building blocks together to create something more complex.

Classes

Classes in C# are blueprints for creating objects. A class can contain fields (variables), methods (functions), and other members. All scripts in Unity are classes that inherit from a base class such as `MonoBehaviour`, `ScriptableObject`, or others to illustrate the characteristics of classes in C#. Let's go through an example class called `XRGrab`, which inherits from the base class, `MonoBehaviour`. The first section of this class consists of the following:

```
public class XRGrab : MonoBehaviour
{
    public XRGrabInteractable grabInteractable;
    private bool isGrabbed = false;
```

In the preceding example, the `XRGrab` class consists of two fields. `public XRGrabInteractable grabInteractable` is a reference to an `XRGrabInteractable` object. This script is typically attached to a GameObject that you want to make interactable in a VR or AR setting; for example, an object the user should be able to grab. `isGrabbed` is a boolean variable that keeps track of whether the interactable object is currently grabbed.

The next section of our class consists of methods. The first method of our class is called `OnEnable()`:

```
    void OnEnable()
    {
```

```
        grabInteractable.onSelectEntered
            .AddListener(Grabbed);
        grabInteractable.onSelectExited
            .AddListener(Released);
    }
```

The `OnEnable()` method adds `Grabbed()` and `Released()` methods as listeners to the `onSelectEntered` and `onSelectExited` events of the `grabInteractable` object. This means when the user interacts with the object in the VR/AR world by selecting it, the `Grabbed()` method will be called, and when the user stops interacting with it and hence deselects it, the `Released()` method will be called.

Let's have a look at what both of these methods might look like:

- Here's the `Grabbed()` method:

```
        void Grabbed(SelectEnterEventArgs args)
        {
            isGrabbed = true;
        }
```

The `Grabbed()` method sets `isGrabbed` to true, indicating that the object has been grabbed.

- Here's the `Released()` method:

```
        void Released(SelectExitEventArgs args)
        {
            isGrabbed = false;
        }
    }
```

The `Released()` method sets `isGrabbed` to false, indicating that the object has been released.

As you can see, scripting in Unity using C# involves defining variables to hold data, implementing functions to manipulate that data or implement gameplay, and organizing these variables and functions into classes that represent objects or concepts in your game. The Unity engine then uses these scripts to drive the behavior of GameObjects within your scenes.

In the realm of C#, class names typically follow the **PascalCase convention**, where each word begins with an uppercase letter and underscores are absent. You have the freedom to select any name that pleases you, as long as it complies with the general naming conventions of C#. Here are the naming guidelines:

- Names should commence with an uppercase letter instead of an underscore or a digit.

- Names can incorporate letters and digits but not underscores.

- Spaces or special characters should not be part of the name.

- Lastly, the name must not be a reserved word in C#. Words such as "class", "int", "void", and so on are reserved and cannot be employed as a name. This is because they have specific meanings in C#, as they are part of its syntax. The compiler expects them to be used in specific ways, so using them as identifiers would cause confusion and lead to compilation errors.

And that's a gentle introduction to scripting in Unity for VR development. Don't worry if it still sounds a bit complex; like with any language, practice makes perfect, and this is exactly what you are going to do in the next section!

Scaling a bus using a slider and C#

We are about to embark on an exciting task: enriching our car exhibition experience by adding a dynamic scaling feature to a bus. Previously, we used fixed animations to scale a police car. However, this time, we will scale a bus in real time based on user interaction with a slider.

With these steps, we will achieve our objective. So, let's dive in:

1. First, we need to bring the bus into our scene. Go to Assets | Simple Vehicle Pack | Prefabs, drag and drop Bus_2 into the scene. Position it at coordinates (0,0,-6), with a -90 degree rotation on the y-axis.

2. Next, we require a slider and a descriptive text that indicates that the bus can be scaled using the slider. To accomplish this, we first need to create a canvas. Right-click in the hierarchy and select **XR | UI Canvas**, then resize it to (0.01,0.01,0), position it at (0,1.5,-5), and rotate it to (0,180,0). Now it is facing the player's direction. Note that you must set the **Render Mode** of the **Canvas** component to **World Space** to manipulate its properties. Let's rename the canvas as Bus Scale Canvas for clarity.

   One thing is missing before adding UI elements to the canvas: the **Vertical Layout Group** component. Let's add this in the Inspector and change the **Child Alignment** property to the **Upper Center**. This ensures that both the slider and text will be vertically aligned at the upper center of the canvas.

3. Next, we can add the slider. Right-click on the **Bus Scale Canvas** and select **UI | Slider**. This will add a slider as a child object of the canvas. The slider consists of several child objects that collectively form the slider's visual and functional features. However, our focus is primarily on the slider object itself. In its **Slider** component in the Inspector, ensure that the **Min Value** is 0 and the **Max Value** is 5, allowing the bus to be scaled between these values. Let's rename the slider to Bus Scale Slider for distinction.

4. For enhanced user experience, we will also place a text above the slider. Add a **Text – TextMeshPro (TMP)** child object to the Slider GameObject by right-clicking on **Bus Scale Canvas** and selecting **UI | Text – TextMeshPro (TMP)**. Download the required TMP assets if prompted. Change **Text Input** to Bus Scale Slider, select **H3** for the **Text Style**, select the **Auto Size** checkbox, and rename it Bus Scale Text.

We have now set up the slider, which serves as our initialization event, as shown in *Figure 5.11*.

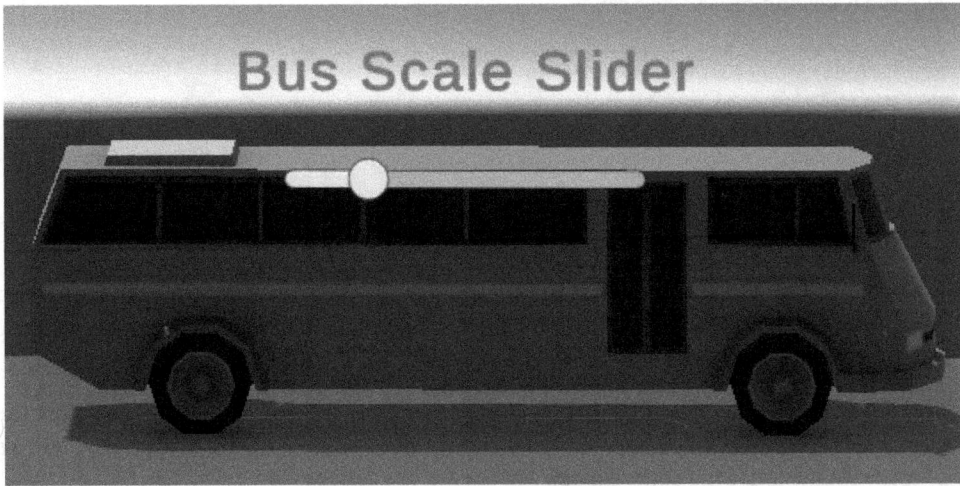

Figure 5.11 – The bus scale slider

Following this, we will link this with the scaling animation and state control via a single C# script. The idea is to bind the slider's value to the bus's scale.

Before proceeding with scripting, it's advisable to maintain organization by creating a new folder:

1. Right-click in the `Assets` folder, choose **Create | Folder**, and name it `Scripts`.

2. Inside this new folder, create a new C# script and rename it `BusScaler`. This script will be our primary tool in bringing dynamic scaling to life.

3. Now double-click on the script. This will open it in the IDE that is installed on your computer. This is the code we are going to develop.

Writing the animation script

The following explanations guide you on how the main components of the `BusScaler` script interact with each other:

1. Start with the following two lines of code to define your namespaces:

```
using UnityEngine;
using UnityEngine.UI;
```

The first step of any C# script in Unity is to include **namespaces**. A namespace is a collection of classes, structs, enums, delegates, and interfaces. This helps organize the code and provides a level of access control. The `using` keyword is used to include namespaces in the script. `UnityEngine` contains all the classes needed for creating games in Unity, while `UnityEngine.UI` contains classes for creating and manipulating UI elements.

2. The class declaration follows next. In our script, let's declare a new class named `BusScaler` with the following code sequence:

    ```
    public class BusScaler : MonoBehaviour
    ```

    This class inherits from `MonoBehaviour`, which is the base class for all Unity scripts.

3. Next, declare the variables of the `BusScaler` class with the following code sequence:

    ```
    public GameObject bus;
    public Slider slider;
    ```

    The `public` keyword means these variables can be accessed from other scripts and can also be set from Unity's **Inspector** window. The `bus` variable will hold the bus object in the scene and the `slider` variable will hold the slider UI object in the scene.

4. Next, let's override the `Awake()` function to set up initial settings and references before the game starts. This includes setting the initial scale of the bus to 1 on the x-, y-, and z-axes. We do this with the following code snippet:

    ```
    private void Awake()
    {
    bus.transform.localScale = new Vector3(1f, 1f, 1f);
    }
    ```

5. Our final step is to override the `Update()` function. Type or paste in the following lines of code:

    ```
    private void Update()
    {
    float scaleValue = slider.value;
    bus.transform.localScale = new Vector3(scaleValue, scaleValue,
    scaleValue);
    }
    ```

    This code retrieves the value of the slider (which ranges between its minimum and maximum values set in Unity's Inspector) and sets the scale of the bus to this value on the x-, y-, and z-axes. This means if the slider's value is 0.5, the bus will be half its original size.

Testing our animation

Now that our C# script is finished, there are only a few steps left to complete until we can test our animation:

1. In our C# script, the class `BusScaler` is derived from `MonoBehaviour`. To utilize this script, it needs to be associated with a GameObject within the Unity editor. To do this, right-click in the hierarchy and select **Create | Empty GameObject**. Rename this new object `Bus Scaler Controller`.

2. Next, you can drag your script and drop it into the **Inspector** panel of this new GameObject. This action will attach your script as a component.

3. Upon doing this, you'll notice that the public fields bus and slider become visible in the script component within the Unity editor. These fields need to be populated with the actual bus GameObject and the slider object this script will interact with. To do this, simply drag and drop these objects into the corresponding field, as shown in *Figure 5.12*.

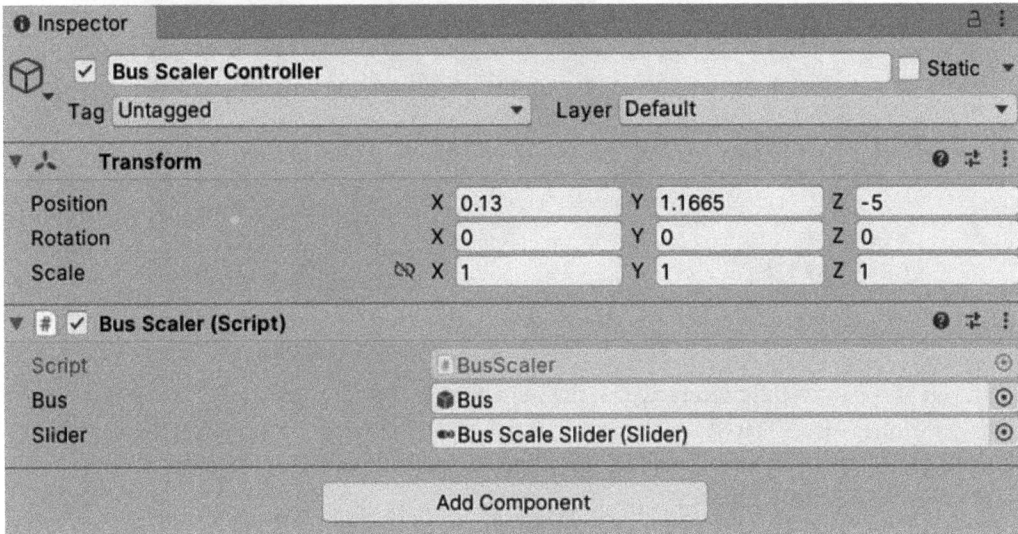

Figure 5.12 – The Inspector window of the Bus Scaler Controller with the bus and slider objects placed into their corresponding fields in the Bus Scaler script

4. The final step before you can test your interactive slider is to place a teleport anchor in front of it. This can be done by going back to our toolbox and navigating to Assets | Samples | XR Interaction Toolkit:

    I.   Here, you'll find the **Teleport** prefab. Drag and drop this to create a third teleport anchor, which we'll call Bus Teleport Anchor.

    II. Position this anchor at coordinates (0,0,-2.5) and adjust its size to (2,1,2). With this setup, the user can now teleport to a position in front of the slider, enabling them to scale the bus interactively, as shown in *Figure 5.13*.

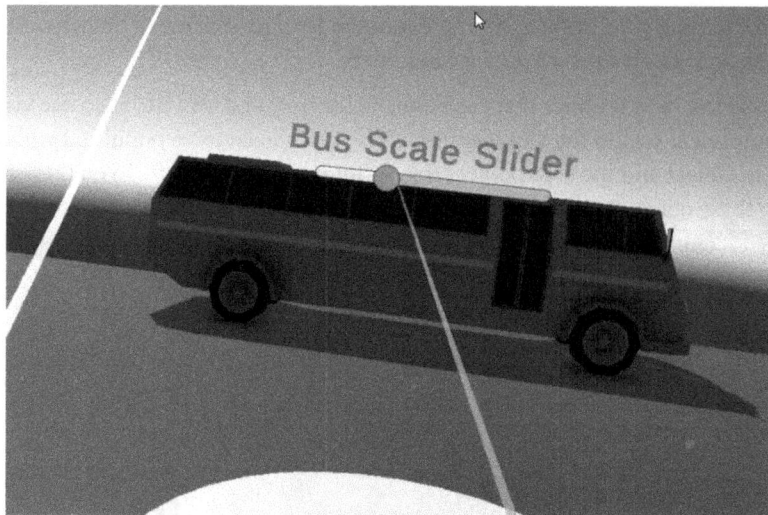

Figure 5.13 – How the bus is scaled when the user interacts with the slider in VR

Hurray, you have successfully completed all the steps needed for this animation. Now, it is time to explore your interactive VR scene once more. For this, follow the steps described in the *Deploying and testing VR experiences onto different VR platforms or simulators* section of *Chapter 3*.

Summary

In this chapter, you have learned how to add interactivity into your VR scenes in Unity, using both code-dependent and code-free methods. By now, having successfully navigated through the steps provided in this chapter, you should have brought your Unity scene to life and acquired a solid understanding of and established a comfort level in deciding when to opt for Unity's animator system and when to leverage the power of C# for your interactive scene creation needs.

We've delved into the intricacies of triggering button events and utilizing the animator system, thereby equipping you with the skills to create simple yet effective interactions in your VR landscape without having to write a line of code. For more intricate interaction designs, you should now feel very familiar with the essential functions of C# in Unity and be familiar with the robust functionalities provided by the XR Interaction Toolkit.

Be it industrial applications or academic projects, these skills and techniques should empower you to create immersive VR scenes in Unity using the XR Interaction Toolkit and its demo scene. You should feel ready to apply what you've learned to your unique use cases.

But our journey does not end here. In the forthcoming chapter, we'll broaden our horizon by venturing into the creation of interactive AR and MR experiences in Unity. This will ensure that you are comprehensively equipped to add interactions across the spectrum of XR scenes. Let's continue our exciting exploration into the world of immersive technology.

6

Building Interactive AR Experiences

Having mastered the basics of crafting and implementing an AR experience in Unity for Android and iOS platforms in *Chapter 4*, we'll now embark on the engaging journey of enhancing your AR scenes with intriguing interactions. This chapter will open a whole new world of user experiences for you as we strive to augment reality with utility-based overlays.

Harnessing the robust capabilities of C#, our adventure will take a turn toward creating an innovative AR application tailored to the food industry. This application will revolutionize the way customers make their selections, offering an unprecedented method for placing orders.

As we unravel the creation process of this AR app, you'll discover how to anchor and scale 3D models in the real world, facilitate user interactions with the elements of your AR app, and comprehend the broad principles to contemplate when architecting your own AR application. This immersive journey will equip you with a rich set of skills, offering a deeper understanding of the boundless possibilities of AR.

This chapter covers the following topics:

- Understanding the design patterns and core components of an interactive AR application
- Building the foundation of our AR menu application
- Adding interactivity to our AR menu application

Technical requirements

To fully participate in and benefit from the AR application development process detailed in this chapter, there are several technical requirements your hardware must meet. If you followed along with the project we built in *Chapter 4*, you can skip these requirements, so long as your Unity setup hasn't changed.

To ensure you can follow along with the content and examples in this book, confirm that your computer system can handle *Unity 2021.3* LTS or a more recent edition, including Android or iOS Build Support.

Additionally, as we delve into the fascinating world of AR, it is helpful to have either an Android or iOS device that can support *ARKit* or *ARCore*. You can check the compatibility of your device at `https://developers.google.com/ar/devices`.

However, if you don't have a device that fits these requirements, you can still participate in these tutorials and test your application on your PC. The *Deploying AR experiences onto mobile devices* section of *Chapter 4* detailed how to do this.

Understanding the design patterns and core components of an interactive AR application

In this section, we will delve into the commonly used design patterns and applications of AR, equipping you with the knowledge to choose the appropriate use case for your AR application. This includes familiarizing yourself with the core concepts and integral components of the AR menu application that we will construct in this chapter.

Understanding the right applications for interactive AR apps is essential before embarking on your journey to develop your first full-fledged AR application. Let's take a look at the various types of AR apps that currently exist, highlighting the different ways they enhance the user experience. Not only will this give you a clearer picture of the current state of AR applications, but it will also shed light on what types of AR experiences resonate most with users.

Let's look at these different kinds of applications:

- **AR gaming apps**: Games such as Pokémon Go and Minecraft Earth are popular examples of AR apps. They place virtual objects such as creatures or blocks into the user's real-world surroundings. Interaction is usually done through touchscreen commands, with the game responding to physical location data from the device's GPS.

- **AR navigation apps**: Apps such as Google Maps Live View and CityViewAR overlay directional cues and points of interest onto real-world images. These apps use GPS data and device orientation to decide which AR elements to show and where. These elements can include arrows, labels, and even 3D models of buildings. Users primarily interact by moving and looking around. Some apps also allow users to touch screen elements for more information.

- **AR shopping apps**: IKEA Place and Amazon AR View let users visualize products in their own homes before buying. These apps access the users' camera feeds so that they can place 3D models of products in their rooms. Advanced apps use environmental understanding to place objects on surfaces and scale them correctly. Users move around to view the product from different angles and use touch commands to select different items, change colors or features, and make a purchase.

- **AR education apps**: Apps such as Star Walk 2 and AR Anatomy Learning are designed to provide immersive, interactive learning experiences. These apps often use image recognition to anchor 3D models such as a star system or a human organ to specific locations in the real world. Users can interact with the 3D models through touch commands, for example, to rotate the model or activate animations. Users may also move around to view the model from different angles.

These different AR applications may cater to different user groups and industries, but they share a core pattern: projecting 3D models into real-world contexts using AR. These apps use the camera feed, device orientation, and GPS data to position and scale the 3D models. User interaction is achieved through touch commands or physical movement, with the interaction often centered on manipulating specific features of the 3D models. This pattern, which emphasizes simplicity and focus, is the current standard in AR app development and will serve as our blueprint for the interactive AR app we'll discuss in this chapter. It's also a pattern we recommend for our own AR projects.

> **Note**
> Remember that the goal is not to force AR functionality into an app but to use it where it enhances user experience. For instance, using AR to display a restaurant menu on a table might seem novel, but it doesn't necessarily add value for the user. A traditional 2D app or physical menu would serve the same purpose. However, if you create an AR experience where users can visualize each dish on the table, scaled to the actual serving size, the user gets a clear sense of what to expect. This enhances the dining experience and overall satisfaction, showing how AR can create meaningful and beneficial interactions when applied appropriately. And this is exactly what we are going to create in the subsequent sections.

With this understanding, let's proceed and build the foundation of our AR menu application.

Building the foundation of our AR menu application

As we start crafting this AR menu application, you'll set the groundwork for your AR project by assembling the essential components. The following sections will guide you through the process of sourcing and importing a variety of dishes from the Unity Asset Store and adding UI buttons and text elements to your scene, providing the basis for your interactive AR menu application.

Defining the key components of our AR application

Throughout this chapter, we will be building an AR application that allows users to place 3D models of food items that have been detected by their mobile device's camera onto real-world surfaces. Before we get into the project, it is always good to visualize the functionality of the app we are going to create. Our AR application will consist of several key components:

- **AR Session and AR Session Origin**: These are the primary components for any AR Foundation application. These form the stage for our performance. For an in-depth understanding of AR Session Origin and AR Session, you're invited to revisit the *Exploring the AR Session Origin GameObject* and *Understanding the AR Session GameObject* sections of *Chapter 4*.

- **AR Raycast Manager and AR Plane Manager**: These components come by default with AR Session Origin. AR Raycast Manager performs raycasts against tracked AR features and geometry, such as detected planes. We'll use it to find where we can place our 3D models in the real world. AR Plane Manager, on the other hand, allows you to detect and track horizontal and vertical surfaces in the real world.

- **Food prefabs**: These are 3D models of food items that users can place in the real world. They are the tangible objects that the user will interact with.

- **UI buttons and text elements**: These UI elements are needed to guide the user on how the app works, to swap the current food prefab to the next or previous one in line, and to display nutritional information for each dish. Consider these UI elements as the interactive elements of our application where the users can participate.

- **The ARPlacePrefab script**: This is our main script, and it's responsible for managing the placement of the 3D models and a UI button in the AR space. It also handles the user interactions for placing objects and updates the placement indicator based on AR raycast results.

- **The SwapPrefab script**: This script works in conjunction with the **ARPlacePrefab** script to switch between different food prefabs when the UI button is clicked.

- **The Food script**: This script defines a class named Food, which enables us to create Food objects with certain characteristics or properties, such as a name, ingredients, calories, and diet type.

By implementing these components, we will have a working AR application that lets users place, view, and swap different 3D models of food in their real-world environment using their smartphone in place of a traditional printed menu. To ensure that the users understand what they are supposed to do in this application, dedicated UI elements will be added to the start scene of the application. By using visual cues such as grids, the process of finding the optimal position to place a dish will be facilitated for the user. By adding a scaling functionality, the user will be able to adjust the size of the displayed dish for an enhanced user experience.

In the following section, we will set up our AR application so that we're one step closer to building an interactive AR app.

Setting up the environment

To initiate our journey, let's start by creating a new project, as explained in the *Creating an AR project with Unity's AR template* section in *Chapter 4*. If you've already completed these steps, you can simply open the existing project for this chapter. After your project loads, follow this step-by-step guide to set up your environment:

1. Let's kick-start the process by crafting a new empty scene. This can be achieved by going to **Assets | Scenes** in the **Project** window, right-clicking, and opting for **Create | Scene**. Name this new scene ARFoodMenu.

2. A quick double-click will transport you into your newly minted scene, which currently only houses **Main Camera** and **Directional Light**. **Main Camera** can be removed as it won't be needed.

3. Before diving into the AR setup, it's important to lay the groundwork for our AR app. In our situation, this means incorporating **AR Session Origin** and **AR Session**. You can accomplish this by right-clicking in the **Scene Hierarchy** window, choosing **XR | AR Session Origin**, and repeating the same step for **AR Session**.

4. Once **AR Session Origin** is in your scene, select it and navigate to the **Inspector** window. Click the **Add component** button, search for the AR Plane Manager script, and select it. Repeat this process for the **AR Raycast Manager** script.

5. In the **Project** window, search for ARPlane and drag the prefab into the **Plane Prefab** cell of the **AR Plane Manager** script component in the **Inspector** window. As we just want to detect horizontal planes for our AR menu application, select **Horizontal** for **Detection Mode**. We don't need to assign any prefabs for **AR Raycast Manager** as we just need the functionality to calculate when the ray encounters the surface.

And there you have it! With a few clicks, we've added **AR Session**, **AR Session Origin**, **AR Raycast Manager**, and **AR Plane Manager** to our scene. Our next move is to import food models from the Asset Store, which will showcase the meals we aim to present to the user.

Designing 3D food models

As a restaurant proprietor, you need to produce 3D models of your meals at this stage. The following are a few strategies you could employ to design 3D models of food or dishes:

- **3D modeling software:** There are many 3D modeling software tools you can use to create 3D models. These include *Blender*, *3ds Max*, *Autodesk Maya*, and *SketchUp*. Among these, *Blender* is a popular choice because it's free and open source. There are many online tutorials available to help beginners get started with *Blender*, such as *CGFastTrack*'s YouTube channel (https://www.youtube.com/@CGFastTrack/featured).

- **Photogrammetry**: Photogrammetry is a technique where you take multiple photographs of an object from different angles and then use software to stitch them together into a 3D model. Tools such as *Luma Labs*, *Reality Capture*, *Meshroom*, and *Agisoft Metashape* can help with this. For food models, this can give a very realistic result, but it may also be more complex to get right, particularly with foods that are uniform in color or texture.

- **Download pre-made 3D models**: If creating models is too time-consuming or outside your current skill set, you can download pre-made models from sites such as *Sketchfab*, *TurboSquid*, or the *Unity Asset Store*, which is what we will do for this project. Just make sure you check the licensing terms to see if they fit your needs.

- **Hire a 3D artist**: If you have a budget for your project, you might consider hiring a 3D artist to create custom food models for you. This can give you exactly what you want, tailored to your specifications.

As we pointed out previously, we are going to import our models through the Unity Asset Store for the sake of this AR menu application. Follow these instructions to do this:

1. Go to the Unity Asset Store (`https://assetstore.unity.com/`) and search for the `French Fries - Free` package. Alternatively, you can access the **French Fries** package page via this link: `https://assetstore.unity.com/packages/3d/props/food/french-fries-free-164017`.

2. After adding the package to your assets, click the **Open in Unity** button that now appears on the Asset Store's website.

3. Your Unity project should open, and you should be prompted with the option to import the package. Click the **Import** button to proceed.

4. Once the package has been successfully imported into your project, repeat the previous three steps for the **Japanese Food Tofu – FREE** package (`https://assetstore.unity.com/packages/3d/props/food/japanese-food-tofu-free-203533`) and the **Japanese Food Kashipan – FREE** package (`https://assetstore.unity.com/packages/3d/props/food/japanese-food-kashipan-free-210938`).

Now that we've imported the appropriate food meals into our scene, we fulfilled the food prefabs requisite. In the next section, we will add the UI button and text components that will guide the user through the AR menu application.

Adding UI buttons and text elements to our AR menu application

In this section, we'll focus on integrating three buttons and two text elements into our scene. These buttons are divided into one for initiating the app and two for transitioning between the dishes. The first text element serves to guide the user through the AR menu application, while the second acts as a placeholder to contain information about the displayed dish, such as its name, ingredients, caloric content, and dietary type.

Excluding the second text element, these UI elements should remain within the user's field of view, regardless of how the device is rotated. This can be accomplished by assigning them as child objects to the AR camera.

Let's start with the two dish-swapping buttons. Follow these steps to achieve this:

1. Navigate to the **Scene Hierarchy** window, right-click on the **AR Camera** component, and select **UI | Canvas**. This action will add **Canvas** as a child object of **AR Camera**. The **Canvas** component is necessary for all UI elements.

2. Alter the Canvas's **Render Mode** to **World Space** in the **Inspector** window. Stay in the **Inspector** window and resize the **Canvas** component for all three directions ($0.01, 0.01, 0.01$) with **Width** set to 480 and **Height** set to 90. Position the **Canvas** component at ($0,0,1$).

3. Create the buttons by right-clicking on the **Canvas** component and selecting **UI | Button TextMeshPro**. Perform this step for both buttons and then rename them ButtonNext and ButtonPrevious. Rescale both buttons to ($0.07, 0.32, 1$) with **Width** set to 160 and **Height** set to 30. Position **ButtonNext** at ($10,40,0$) and **ButtonPrevious** at ($-10, 40, 0$).

4. For the child objects of both buttons, enter arrows such as - > and < - as the text input in the **Inspector** window. These arrows tell the user which button switches to the next or previous dish.

At this point, the two buttons have been fully initialized. Next, we need to create an instructional piece of text to guide users on how to use the AR menu application. To achieve this, follow these step-by-step instructions:

1. Right-click on **Canvas**, select **UI | Panel**, and scale it to ($0.17, 0.17, 0.17$). Adjust the panel top to -46.5 and the panel bottom to -3.4. This panel will act as a backdrop for our text. You're free to choose its color; in our case, we chose red. Rename this panel InfoPanel.

2. For the instructional text, right-click **InfoPanel** in the **Scene Hierarchy** window, select **UI | Text TextMeshPro**, and rename the text to InstructionText. Adjust its scale to ($0.7, 0.7, 0.7$) with **Width** set to 600 and **Height** set to 150; then, position it at ($0, 0, 0$).

3. Input the following text: Wait until the blue surface is visible. This shows the area where the food is placed. This instruction will be critical for the user as we'll implement a blue grid area that indicates where the food prefabs will be placed.

4. To prevent the user seeing a meal directly when the application starts, create another button below the instruction text. This button is responsible for spawning the first dish. To do this, right-click **InfoPanel** and select **UI | Button TextMeshPro**. Rename this button PlaceFirstMealButton, scale it to ($2.2, 2.2, 2.2$), and position it at ($0, -292, 0$) with **Width** set to 160 and **Height** set to 30.

5. Open the child text component of **PlaceFirstMealButton**, select it, and enter Place First Meal as the text input in the **Inspector** window. Activate the **Active auto size** checkbox to make sure that the text is adjusted with the button. Both **InstructionText** and **PlaceFirstMealButton** are child objects of **InfoPanel**. They are designed to disappear after the first meal is placed. This means we only need to target **InfoPanel** in our code instead of all three objects individually.

Lastly, we require another text element, but this time, it should not be a child object of **AR camera**. This text needs to spawn above the dishes and remain there consistently.

To accomplish this, let's perform the following steps:

1. Create another **Canvas** component by right-clicking in the **Scene Hierarchy** window and selecting **UI | Canvas**. Scale **Canvas** to (0.1, 0.1, 0.1) and position it at (0.3, 0.3, 1) with **Width** set to 480 and **Height** set to 90. Rename this **Canvas** to FoodInfoCanvas.

2. Right-click **FoodInfoCanvas** and select **UI | Text TextMeshPro**. Rename this text element FoodInfoText, scale it to (0.005, 0.005, 0.005), and position it at (-3.4, 0,31, 0) with **Width** set to 200 and **Height** set to 50. **FoodInfoText** does not need text input as this will be assigned programmatically in correspondence with the displayed dish.

At this point, your **Scene Hierarchy** window should look like what is shown in *Figure 6.1*.

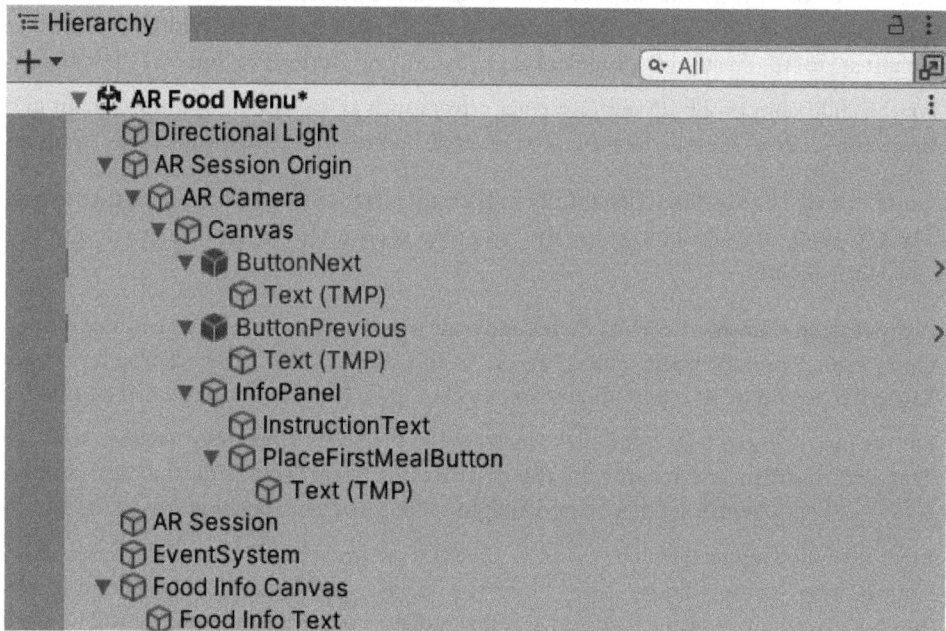

Figure 6.1 – Our project's Scene Hierarchy window, showcasing all components in our scene

Similarly, all your UI elements should be aligned in your Game view, as shown in *Figure 6.2*.

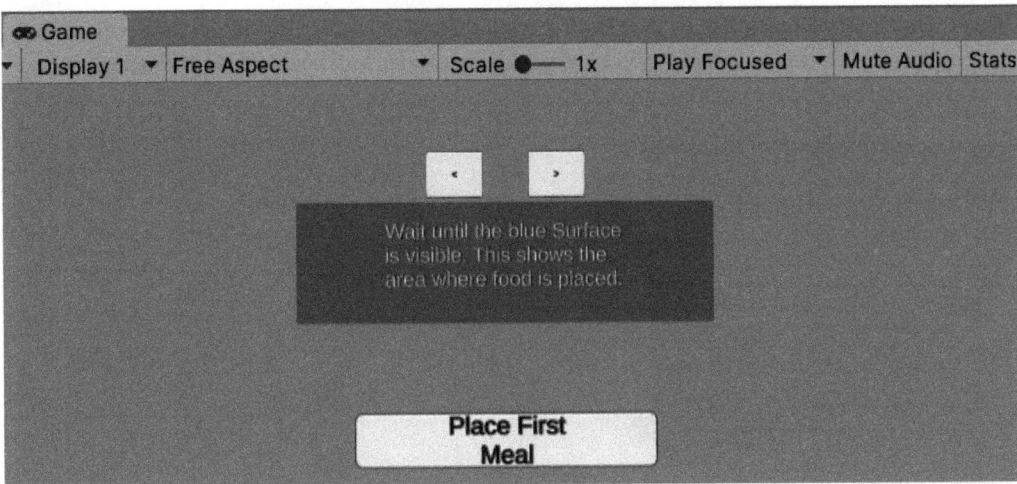

Figure 6.2 – Our project's Game view, showcasing the alignment of the different UI elements in the scene

Please note that all UI elements are positioned and scaled concerning the **Canvas** component. If your UI elements have a different alignment to the one shown in *Figure 6.2*, you can manually adjust the position and scale of your **Canvas** component.

With that, we have successfully added all the necessary UI elements to our scene and therefore finished adding UI buttons and text elements to create an interactive AR application. Next, we'll add scripts to our scene to introduce some interactivity and life to it.

Adding interactivity to our AR menu application

To infuse life into our currently modest AR scene, we must incorporate some C# scripts into our project. This key step will empower the users of our AR app, allowing them to interact with our imported food prefabs dynamically and engagingly.

In the next section, you will get an overview of the short but decisive **Food** script.

Adding the Food script to the scene

Before we dive into scripting, let's first organize our project workspace so that we can work more efficiently by finding the needed components quickly.

Start by adding a new folder to the Assets directory in the **Project** window. Simply right-click and select **Create | Folder**. Rename this folder Scripts and enter it. Right-click in this empty folder and select **Create | C# Script** to create a script called Food. After double-clicking on this script, it should be opened in your preferred **integrated development environment** (**IDE**).

The **Food** script defines a custom data type or class named **Food**. You can think of a class as a blueprint for creating objects. In this case, any Food objects that are created will have certain characteristics or properties, just like how every car has a make, model, and color.

Here's a simple breakdown of this script:

```
[System.Serializable]
public class Food
{
    public string name;
    public string ingredients;
    public string calories;
    public string dietType;
    public GameObject prefab;
}
```

[System.Serializable] is a directive that tells Unity to make this class serializable. This means that Unity will be able to save and load Food objects and that you'll be able to edit Food objects directly in the Unity editor.

After defining a new, public class called Food, a public property called name is declared. This property represents the name of the food, such as French Fries. Next, we declare additional properties such as ingredients, calories, and dietType. The latter property stores the type of diet the food fits into, such as Vegetarian, Vegan, or Paleo.

Lastly, a public property that will hold a GameObject is declared that represents the 3D model of the food that will be displayed in the AR environment.

As we venture into the next section, we'll continue our journey by incorporating the most pivotal C# script in our scene – the **ARPlacePrefab** script. It is this crucial component that will truly open the doors of interactivity within our AR menu application.

Adding the ARPlacePrefab script to the scene

Similar to the previous scripts we created in this chapter, navigate to the Assets folder in the **Project** window of the Unity editor. Now, right-click in this folder and select **Create | C# Script**. This will add a new script to your scene. Change its name from **NewBehaviourScript** to ARPlacePrefab.

> **Note**
>
> We highly recommend that you clone the *Chapter 6* project, which can be found in the GitHub repository of this book, and open it in parallel, enabling you to revisit and mimic the steps described in this project tutorial more easily.

We can open the just-created ARPlacePrefab script by double-clicking on it.

> **Tip**
> The script will launch in the default IDE that you usually utilize. If you wish to switch to a different IDE, simply go to **Edit | Preferences** in the Unity editor. In the newly opened **Preferences** window, navigate to the **External Tools** tab and alter **External Script Editor** to the IDE of your preference.

The `ARPlacePrefab` script, as previously highlighted, will serve as our primary script. It carries out the crucial role of handling the placement of the 3D models and UI buttons within the AR space. Moreover, it manages user interactions so that it can position objects and dynamically update the placement indicator based on AR raycast outcomes.

Next, let's take a look at the various methods and functions in the `ARPlacePrefab` script.

Understanding the ARPlacePrefab script

As the `ARPlacePrefab` script is the core of our AR menu application, we must understand every method of it to truly grasp the mechanisms of interactive AR experiences and successfully replicate the script's functionality in a wide range of projects and contexts.

In the following code snippets, you'll notice some lines that have been commented out. This is because they reference the **SwabPrefab** script, which hasn't been created yet, and would result in compile errors if left uncommented.

At the top of the `ARPlacePrefab` script, various public and private variables are declared:

```
public class ARPlacePrefab : MonoBehaviour
{
    public GameObject ObjectToPlace;
    //public SwapPrefab SwapPrefabScript;
    public GameObject NextPrefabButton;
    public GameObject PreviousPrefabButton;
    public TextMeshProUGUI InfoText;
    public Button PlaceFirstMealButton;
    public GameObject InfoPanel;

    private ARRaycastManager _arRaycastManager;
    private Pose _placementPose;
    private bool _placementPoseIsValid = false;
    private GameObject _placedObject;
    private GameObject _nextButton;
    private GameObject _previousButton;
    private Vector2 _oldTouchDistance;
    private GameObject _placementIndicator;
    private GameObject _placementGrid;
```

The public variables of the ARPlacePrefab script allow users to interact with the AR space, such as positioning ObjectToPlace or cycling between 3D models using NextPrefabButton and PreviousPrefabButton. They also enable users to view details via InfoText and InfoPanel, as well as activate specific AR placements with PlaceFirstMealButton. Additionally, they provide a space for extended details or options.

On the other hand, the private variables govern the internal mechanics of the AR experience. _arRaycastManager handles surface detection in the real-world environment, while _placementPose denotes the desired position for the AR object. Its validity is tracked by _placementPoseIsValid. Once an object is placed in the AR space, it's referred to as _placedObject. Touch gestures utilize _oldTouchDistance. Lastly, _placementIndicator and _placementGrid offer visual cues to users, indicating where an AR object will be positioned.

> **Important note**
>
> Some lines, related to the yet-to-be-introduced **SwapPrefab** script, have been commented out to prevent any compilation issues. We will uncomment them toward the end of our scripting process.

The first method of the ARPlacePrefab script is the Start() method:

```
void Start()
    {
        _arRaycastManager =
            FindObjectOfType<ARRaycastManager>();
        //_nextButton = InstantiateButton(NextPrefabButton,
        //SwapPrefabScript.SwapFoodPrefab);
        //_previousButton =
        //InstantiateButton(PreviousPrefabButton,
        //SwapPrefabScript.SwapToPreviousFoodPrefab);
        _placementGrid = CreatePlacementGrid();
        PlaceFirstMealButton.onClick
            .AddListener(PlaceObjectIfNeeded);
    }
```

This function is called when the program first starts. It's where the initial setup for the script is done, such as finding the **ARRaycastManager** component, creating instances of the next and previous buttons, and setting up a grid for object placement. The lines that define _nextButton and _previousButton have been commented out because they reference the yet-to-be-created **SwapPrefab** script. This script is associated with the two buttons that enable users to cycle through various food choices.

The next method that is called in the ARPlacePrefab class is the Update() method:

```
void Update()
    {
        if (Input.touchCount == 2)
```

```
    {
        PinchToScale();
    }

    UpdatePlacementPose();
    UpdatePlacementIndicator();

    if (_placementPoseIsValid && Input.touchCount > 0
        && Input.GetTouch(0).phase == TouchPhase.Began)
    {
        PlaceObject();
    }
}
```

The Update() method is like the heartbeat of the script, constantly checking on things every moment the game is running. If a two-finger touch input is detected, it adjusts the scale of the placed object based on how much the distance between the fingers changes, similar to a pinch or stretch movement. It also updates the placement pose and indicator, which determines where the object will be placed, and listens for a single touch input. If a single touch input is detected, it places the object.

The PlaceObject() method follows after the Update() method and looks like this:

```
public void PlaceObject()
    {
        if (_placedObject != null)
        {
            Destroy(_placedObject);
        }

        _placedObject = Instantiate(ObjectToPlace,
            _placementPose.position,
            _placementPose.rotation);
        _placedObject.transform.localScale = new
            Vector3(0.5f, 0.5f, 0.5f);
        PositionButton(_nextButton, new Vector3(0.5f, 0f,
            0f));
        PositionButton(_previousButton, new Vector3(-0.5f,
            0f, 0f));

        UpdateFoodInfoText();

        _placementGrid.SetActive(false);
    }
```

This function is called when a single touch input is detected in the `Update()` function. It places the selected object in the real world at the location determined by the placement pose. It also positions and activates the next and previous buttons. Then, it fetches some information about the currently selected food item and updates the `InfoText` UI with this information.

The next method in the `ARPlacePrefab` class is the `UpdatePlacementPose()` method:

```
private void UpdatePlacementPose()
{
    var screenCenter =
        Camera.current.ViewportToScreenPoint(
        new Vector3(0.5f, 0.5f));
    var hits = new List<ARRaycastHit>();
    _arRaycastManager.Raycast(screenCenter, hits,
        TrackableType.Planes);

    _placementPoseIsValid = hits.Count > 0;
    if (_placementPoseIsValid)
    {
        _placementPose = hits[0].pose;
        PositionGrid(hits[0].pose.position,
            hits[0].pose.rotation);
    }
    else
    {
        _placementGrid.SetActive(false);
    }
}
```

This method is called in the `Update()` function. It acts like the script's eyes in the sense that it's always looking at the middle of your screen and figuring out where in the real world that corresponds to in the virtual world. This method updates the placement pose, which is the position and orientation where the new object will be placed. It does this by casting a ray from the center of the screen and checking if it hits a plane in the AR environment.

The `UpdatePlacementIndicator()` method works similarly to the `UpdatePlacementPose()` method and looks like this:

```
private void UpdatePlacementIndicator()
{
    if (_placementPoseIsValid)
    {
        _placementIndicator.SetActive(true);
        _placementIndicator.transform
            .SetPositionAndRotation(
```

```
                _placementPose.position,
                _placementPose.rotation);
        }
        else
        {
            _placementIndicator.SetActive(false);
        }
    }
```

This function is also called in the `Update()` function. It simply moves and rotates the placement indicator to match the placement pose, showing the user where the object will be placed. It also turns `LineRenderer` on or off, depending on whether a suitable location for placing the object was found.

The next method is called `InstantiateButton()` and looks like this:

```
private GameObject InstantiateButton(GameObject
buttonPrefab, UnityEngine.Events.UnityAction onClickAction)
    {
        var button = Instantiate(buttonPrefab);
        button.SetActive(false);
        button.GetComponent<Button>().onClick
            .AddListener(onClickAction);
        return button;
    }
```

The `InstantiateButton` method serves as a vital cog in the `ARPlacePrefab` class. Its primary role is to create and set up the buttons that users interact with to navigate between food choices. When invoked, the method takes in two parameters: a button prefab and an action to be executed upon clicking the button.

The method begins by creating a new instance of the button using the provided prefab. Once instantiated, the button is initially set to be inactive, ensuring it doesn't immediately appear or interfere with the user interface. Then, the method fetches the button's built-in **On Click** event and attaches the specified action to it, ensuring the desired function is executed when the button is pressed. Finally, the newly created and configured button is returned by the method, ready for use in the AR interface.

The next two methods in the script act as helper functions. They are called `CreatePlacementGrid()` and `PlaceObjectIfNeeded()`:

```
private GameObject CreatePlacementGrid()
    {
        var grid =
            GameObject.CreatePrimitive(PrimitiveType.Plane);
        grid.transform.localScale = new Vector3(0.01f,
            0.01f, 0.01f);
```

```
        grid.GetComponent<Renderer>().material =
            Resources.Load<Material>("GridMaterial");
        grid.SetActive(false);
        return grid;
    }


    private void PlaceObjectIfNeeded()
    {
        if (_placementPoseIsValid)
        {
            PlaceObject();
            InfoPanel.SetActive(false);
        }
    }
}
```

The `CreatePlacementGrid()` method creates a little grid in the virtual world to help you see where your dish will be placed before you place it. If everything has been set up correctly, the `PlaceObjectIfNeeded()` method places the food prefab in the virtual world and then hides `InfoPanel`, cleaning up your screen. Pay attention to the following line in the `CreatePlacementGrid()` method:

```
grid.GetComponent<Renderer>().material =
Resources.Load<Material>("GridMaterial");
```

Here, we assign a **GridMaterial** component to `grid` so that the user sees a visual cue of where the food prefab will be placed. Besides using the Unity editor's interface, C# scripting provides us with another way to assign materials to `GameObject` – that is, by defining the precise path and name of the material that we want to use. In this case, a material named `GridMaterial` is expected to be in the `Resources` folder of our project.

Before continuing with the next few methods, let's quickly create a material called `GridMaterial` in the `Resources` folder of our project. Simply head to the **Project** window, right-click inside the `Assets` folder, and select **Create | Folder**. Rename this folder `Resources` and enter it. Right-click inside this folder and select **Create | Material**. Rename the material `GridMaterial` and select a blue albedo color of your choice.

Now, back in the `ARPlacePrefab` class, let's have a look at the `PinchToScale()` and `ScalePlacedObject()` methods, which look like this:

```
private void PinchToScale()
    {
        Touch touchZero = Input.GetTouch(0);
        Touch touchOne = Input.GetTouch(1);

        if (touchZero.phase ==
```

```
        TouchPhase.Moved || touchOne.phase ==
        TouchPhase.Moved)
        {
            Vector2 touchDistance =
                touchOne.position - touchZero.position;
            float pinchDistanceChange =
                touchDistance.magnitude -
                _oldTouchDistance.magnitude;

            float pinchToScaleSensitivity = 0.001f;

            if (_placedObject != null)
            {
                ScalePlacedObject(pinchDistanceChange,
                    pinchToScaleSensitivity);
            }

            _oldTouchDistance = touchDistance;
        }
    }

private void ScalePlacedObject(float pinchDistanceChange, float
pinchToScaleSensitivity)
    {
        Vector3 newScale =
            _placedObject.transform.localScale + new
            Vector3(pinchDistanceChange,
            pinchDistanceChange, pinchDistanceChange) *
            pinchToScaleSensitivity;
        newScale = Vector3.Max(newScale, new Vector3(0.1f,
            0.1f, 0.1f));
        newScale = Vector3.Min(newScale, new Vector3(10f,
            10f, 10f));
        _placedObject.transform.localScale = newScale;
    }
```

As the names of these two methods already suggest, they enable the zoom feature of the food prefabs. The `PinchToScale()` method checks whether you're using two fingers to pinch the screen, and if you are, it changes the size of your placed object by calling the `ScalePlacedObject()` method. This functionality is very useful in the context of our AR menu application as it enables the user to scale the displayed food prefab to the desired size.

The `PositionButton()` method provides another useful functionality for application users. It consists of these components:

```
private void PositionButton(GameObject button,
Vector3 offset)
    {
        button.transform.position =
            _placedObject.transform.position + offset;
        button.SetActive(true);
    }
```

The `PositionButton()` method ensures the next and previous buttons are always near your placed object so that you can easily press them. Hence, this method is fundamental to ensure a satisfying user experience of the AR menu application across different devices.

The `UpdateFoodInfoText()` method is another integral part of the `ARPlacePrefab` class. Let's see what the method looks like:

```
private void UpdateFoodInfoText()
    {
        /*Food currentFood =
            SwapPrefabScript.GetCurrentFood();
        InfoText.text = $"<b>Name:</b>
        {currentFood.name}\n<b>Ingredients:</b>
        {currentFood.ingredients}\n
        <b><color=red>Calories:</color></b>
        {currentFood.calories}\n<b>Diet Type:</b>
        {currentFood.dietType}";*/
        InfoText.transform.position =
            _placedObject.transform.position + new
            Vector3(-0.2f, 0.3f, 0f);
        InfoText.transform.rotation =
            _placedObject.transform.rotation;
    }
```

Once you've placed a food prefab in the environment, the `UpdateFoodInfoText()` method tells you more about this specific dish, such as its name, ingredients, how many calories it has, and which diet it aligns with. As before, the initial lines have been commented out due to their reference to the yet-to-be-created **SwapPrefab** script.

Finally, there is only one small script left for you to understand in the `ARPlacePrefab` class. It's the `PositionGrid()` method and it looks like this:

```
private void PositionGrid(Vector3 position,
Quaternion rotation)
    {
        _placementGrid.transform.SetPositionAndRotation(
            position, rotation);
        _placementGrid.SetActive(true);
    }
```

This method simply moves the placement grid to the right spot in the virtual world so that it matches where your indicator is pointing in the real world.

With that, you have successfully understood the `ARPlacePrefab` script, which is the primary script that's used in our AR menu application. In the next section, you will learn how to attach this script to **AR Session Origin** as a component.

Assigning the ARPlacePrefab script as a component

By attaching the main script of any AR application to **AR Session Origin**, we ensure proper alignment with the real world, leverage core AR functionalities, follow common design patterns, and help with the overall performance and coherence of the AR app. This is why we need to assign the `ARPlacePrefab` script as a component of **AR Session Origin** in the **Inspector** window in the Unity editor before progressing to the next step. To do this, simply follow these steps:

1. Select **AR Session Origin** in the **Scene Hierarchy** window and navigate to the **Inspector** window. There, click on the **Add Component** button and search for the `ARPlacePrefab` script. By selecting it, the script's public fields are shown in the **Inspector** window.

2. For the **Object to Place** field, you can assign one of the food prefabs, such as **Tofu**, by dragging and dropping it from the **Project** window into the corresponding cell.

3. Following the same principle, you can drag and drop **ButtonNext** into the **Next Prefab Button** cell, **ButtonPrevious** into the **Previous Prefab Button** cell, **FoodInfoText** into the **InfoText** cell, **PlaceFirstMealButton** into the **Place First Meal Button** cell, and **InfoPanel** into the **Info Panel** cell.

> **Important note**
> The **SwapPrefab** cell expects the **SwapPrefab** script, which we are going to create and assign in the upcoming sections.

Congratulations – you have successfully fulfilled the `ARPlacePrefab` script requirement on our roadmap to create an interactive AR menu application! In the next section, you will come even closer to the final application by delving into the script that swaps our food prefabs: the **SwapPrefab** script.

Adding the SwapPrefab script to the scene

To add the **SwapPrefab** script to your project, navigate to the `Assets` folder in the **Project** window of the Unity editor. Right-click in this folder and select **Create | C# Script**. Rename the newly created script `SwapPrefab`. Double-click on the script to open it in your preferred IDE.

The `SwapPrefab` script is tied to the two buttons that allow the user to cycle through different food options. It starts like this:

```
public class SwapPrefab : MonoBehaviour
{
    public Food[] AvailableFoods;
    private int CurrentFoodIndex = 0;
    private ARPlacePrefab ARPrefabPlacement;
```

In the preceding code snippet, you can see the declaration of a list of `Food` objects called `AvailableFoods`, a variable to keep track of the current food called `currentFoodIndex`, and a reference to the `ARPlacePrefab` script called `ARPlacePrefab`.

Next, let's have a look at the methods of the `SwapPrefab` class to understand how the user can cycle through the different dishes.

The first method that is relevant for this is the `Start()` method. It looks like this:

```
void Start()
    {
        ARPrefabPlacement =
            FindObjectOfType<ARPlacePrefab>();
        if (AvailableFoods.Length > 0)
        {
            ARPrefabPlacement.ObjectToPlace =
                AvailableFoods[0].prefab;
        }
    }
```

This method is called when the scene starts. It grabs a reference to the `ARPlacePrefab` script and sets the first food object in the list as the one to be placed in the AR environment.

Next, there is the `SwapFoodPrefab()` method, which looks like this:

```
public void SwapFoodPrefab()
    {
        CurrentFoodIndex = (CurrentFoodIndex + 1) %
            AvailableFoods.Length;
        ARPrefabPlacement.ObjectToPlace =
            AvailableFoods[CurrentFoodIndex].prefab;
```

```
    ARPrefabPlacement.PlaceObject();

    // Update the InfoText
    ARPrefabPlacement.InfoText.text = $"<b>Name:</b>
    {AvailableFoods[CurrentFoodIndex].name}\n
    <b>Ingredients:</b>
    {AvailableFoods[CurrentFoodIndex].ingredients}\n
    <b><color=red>Calories:</color></b>
    {AvailableFoods[CurrentFoodIndex].calories}\n
    <b>Diet Type:</b>
    {AvailableFoods[CurrentFoodIndex].dietType}";
}
```

This method is linked to a button. When the button is clicked, it advances `currentFoodIndex` to the next food in the list or back to the start if it's at the end and updates `objectToPlace` in the `ARPlacePrefab` script to the new food. Then, it immediately places the new food object in the AR environment and updates the displayed information about the food.

The `SwapToPreviousFoodPrefab()` method in the `SwapPrefab` class consists of the following lines of code:

```
public void SwapToPreviousFoodPrefab()
    {
        CurrentFoodIndex--;
        if (CurrentFoodIndex < 0)
        {
            CurrentFoodIndex = AvailableFoods.Length - 1;
        }
        ARPrefabPlacement.ObjectToPlace =
            AvailableFoods[CurrentFoodIndex].prefab;
        ARPrefabPlacement.PlaceObject();

        // Update the InfoText
        ARPrefabPlacement.InfoText.text = $"<b>Name:</b>
        {AvailableFoods[CurrentFoodIndex].name}\n
        <b>Ingredients:</b>
        {AvailableFoods[CurrentFoodIndex].ingredients}\n
        <b><color=red>Calories:</color></b>
        {AvailableFoods[CurrentFoodIndex].calories}\n
        <b>Diet Type:</b>
        {AvailableFoods[CurrentFoodIndex].dietType}";
    }
```

This method works similarly to `SwapFoodPrefab()`, but it goes back to the previous food in the list instead of moving forward. If it's already at the start of the list, it loops back to the end. Like `SwapFoodPrefab()`, it then immediately places the new food object and updates the displayed information.

Lastly, there is the short `GetCurrentFood()` method:

```
public Food GetCurrentFood()
    {
        return AvailableFoods[CurrentFoodIndex];
    }
```

This is a simple helper method that other parts of the program can call to find out what the current food is.

Remember those lines we commented out in the `ARPlacePrefab` script because they referenced the `SwapPrefab` script? Now is the time to go back to the `ARPlacePrefab` script and uncomment those lines. Once you've done this, we can dive into the last missing snippets of our AR menu application.

Setting up On Click events for the buttons

Now, we need to link **ButtonNext** and **ButtonPrevious** to the `SwapPrefab` script by adding **On Click** events to both. Follow these steps to do this:

1. We need to attach the `SwapPrefab` script to a `GameObject` object in our scene. We can do this by creating an empty `GameObject` object in the **Scene Hierarchy** window of the Unity editor. Rename this `GameObject` to `PrefabManager` and attach the `SwapPrefab` script to it.

2. Now, we must link **ButtonNext** and **ButtonPrevious** to the `SwapPrefab` script. Let's start with **ButtonNext**. Select it in the **Scene Hierarchy** window and navigate to its **On Click** event section in the **Inspector** window. Drag the `PrefabManager GameObject` into the object cell. In the drop-down menu, select the `SwapPrefab | SwapFoodPrefab()` function. This calls the `SwapFoodPrefab()` function of the `SwapPrefab` script, whenever **ButtonNext** is pressed. *Figure 6.3* shows what the **On Click** event of **ButtonNext** should look like.

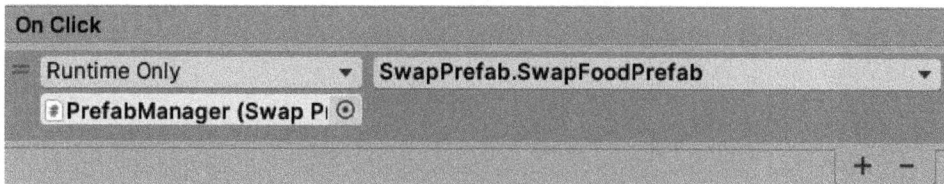

Figure 6.3 – ButtonNext's On Click event

3. Repeat this process for **ButtonPrevious**, but this time, select the `SwapPrefab | SwapToPreviousFoodPrefab()` function from the dropdown.

Do you remember the takeaway from the *Assigning the ARPlacePrefab script as a component* section in that the SwapPrefab script is the last missing cell input of the ARPlacePrefab script that is attached to **AR Session Origin**? Now, the time has finally come to assign the SwapPrefab script to this cell. You can do this in a few, simple steps:

1. Select **AR Session Origin** in the **Scene Hierarchy** window.

2. Search for SwapPrefab via the search bar of the **Project** window.

3. Now, drag and drop the script into the respective cell of the AR Session Origin's ARPlacePrefab script component in the **Inspector** window. *Figure 6.4* shows what your ARPlacePrefab script component should look like now.

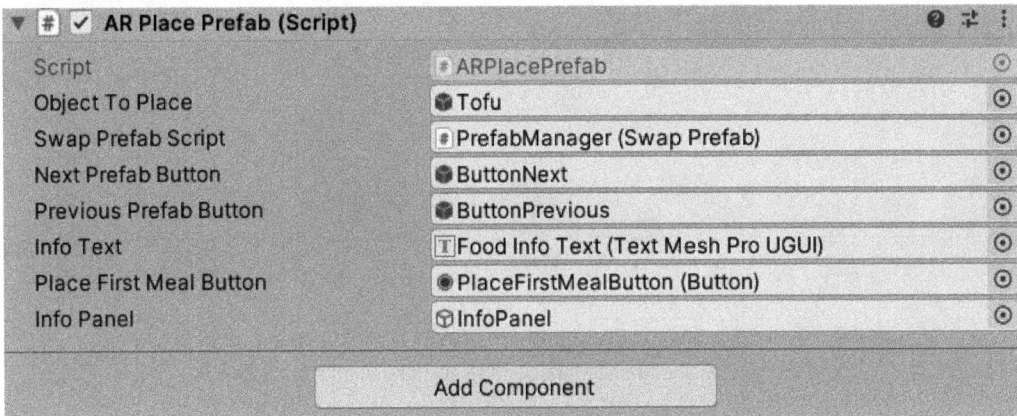

▼ # ✓ **AR Place Prefab (Script)**		❓ ⇄ ⋮
Script	# ARPlacePrefab	⊙
Object To Place	🔶 Tofu	⊙
Swap Prefab Script	# PrefabManager (Swap Prefab)	⊙
Next Prefab Button	⬤ ButtonNext	⊙
Previous Prefab Button	⬤ ButtonPrevious	⊙
Info Text	T Food Info Text (Text Mesh Pro UGUI)	⊙
Place First Meal Button	⬤ PlaceFirstMealButton (Button)	⊙
Info Panel	⬕ InfoPanel	⊙
	Add Component	

Figure 6.4 – The ARPlacePrefab script component of AR Session
Origin once all of its cells have been correctly assigned

Before testing the scene on your PC or deploying it to your phone, there is one last step you have to take: add some nutritional information to our food prefabs. This can be done very easily:

1. Select **PrefabManager** in the **Scene Hierarchy** window and navigate to the **Inspector** window.

2. In the SwapPrefab script component, click the + button three times to add three dishes to our AR menu application.

3. Insert the names, nutritional information, diet type, and prefabs of our three dishes, namely **French Fries**, **Tofu**, and **Kashipan**, into the respective cells, as shown in *Figure 6.5*.

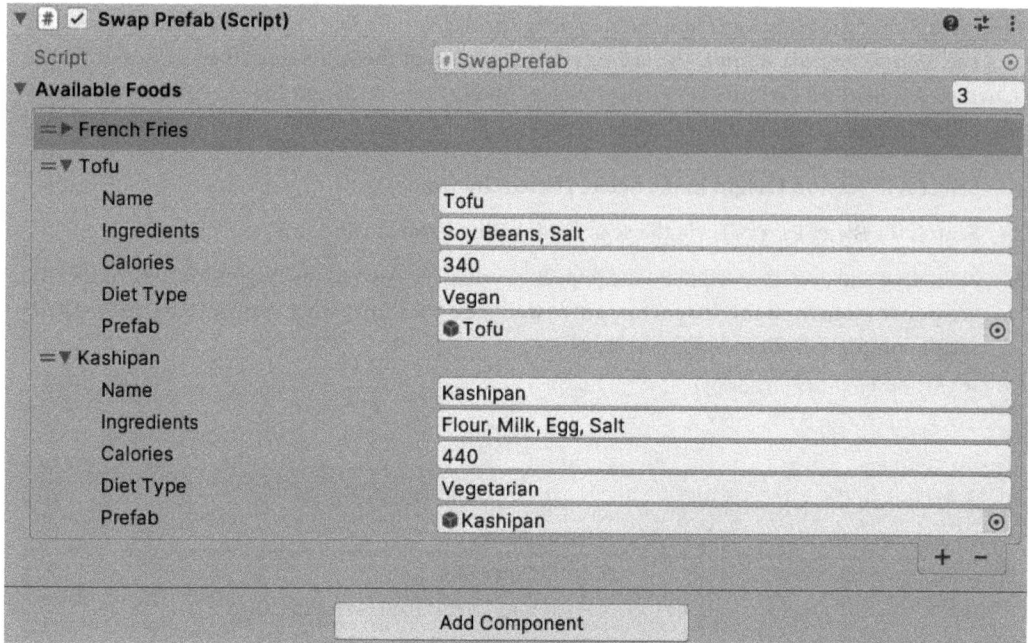

Figure 6.5 – The inserted names, nutritional information, diet type, and prefabs of our three dishes in the PrefabManager's SwapPrefab script component

With that, you have completed the final step in our roadmap to create an interactive AR menu application. To celebrate this milestone, the next section explains how you can explore the different features of your newly developed AR application on your mobile device.

Exploring the completed, interactive AR menu application on a mobile device

To test your interactive AR menu application on your PC or deploy it onto your mobile device, please refer to the step-by-step instructions provided in the *Deploying AR experiences onto mobile devices* section of *Chapter 4*. *Figure 6.6* shows you what should be displayed on your screen once you have successfully deployed the AR menu application onto your mobile device and opened it.

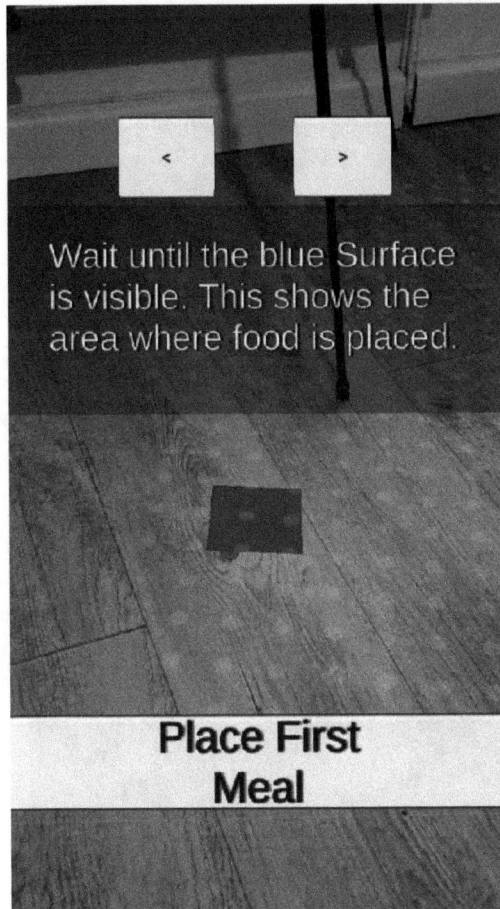

Figure 6.6 – The initial screen of the newly deployed AR menu application on a mobile device

Notice how we can see the instruction text, which familiarizes the user with the nature of the application, and the button at the bottom of the application, which the user has to press to place the first dish on the detected plane. You can also see the two buttons that the user can press to see the next or previous prefab. If there is a horizontal plane in your scene, you will see the blue placement grid element in the center of your screen.

Figure 6.7 shows what the screen of your device should look like once you have pressed the **Place First Meal** button.

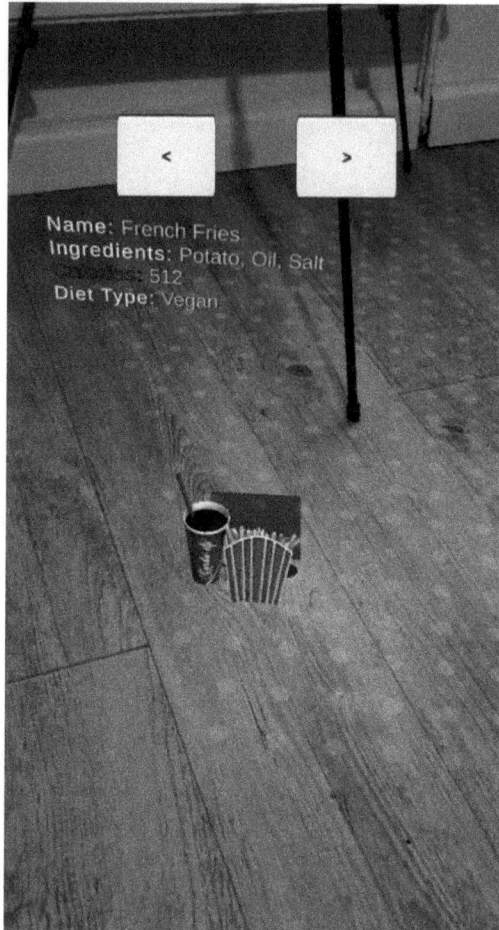

Figure 6.7 – The screen of the AR menu application on a mobile device
once the "Place First Meal" button has been pressed

By touching the screen with two of your fingers and moving them closer together or further apart, you can scale the **French Fries** to become bigger or smaller. By changing the rotation of your mobile device after the first food prefab has been placed and clicking on the -> button, the next food prefab should be placed at the new position, as indicated by the blue placement grid element.

The upcoming section offers a summary of the valuable skills you've acquired in your journey to create XR applications in Unity. Specifically, you've harnessed the power of C# scripting to craft interactive AR applications. This newfound expertise marks a significant enhancement to your repertoire, expanding your capacity to develop sophisticated XR experiences.

Summary

Throughout this chapter, you've embarked on the exciting journey of building your very first interactive AR application, testing it meticulously on your mobile device. This experience should have empowered you to evaluate prospective AR app ideas critically in terms of their utility.

Leveraging intuitive design patterns, you should now feel equipped to create engaging AR applications that allow users to interact seamlessly with virtual objects within a real-world context. From establishing the core components of the AR app, such as user interface and text elements, to invoking their interactivity through C# scripting, you're now well-equipped to create a diverse array of interactive AR applications.

These applications will breathe life into virtual objects placed within real-world environments, inviting users to interact with them in significant and captivating ways. Congratulations on reaching this milestone in your AR development journey. You are now more than capable of bringing your creative AR ideas to life.

Looking forward, the next chapter promises to elevate your VR development skills to a new level. It's all about adding magic to your VR scenes through audio and visual effects. The upcoming chapter will delve into the theory of audio in Unity, guide you through the process of incorporating audio sources and mixers, and unveil the secrets of adding particle effects and animations to your VR experiences. The road ahead is set to be an exciting journey of discovery and application.

7

Adding Sound and Visual Effects

Creating immersive XR experiences requires more than just importing 3D assets and scripting interactivity. For a truly natural and captivating experience, we need to integrate sound and visual effects. Imagine an AR monster hunt game. You've built the logic, but without sound and visual effects, the blend of real world and virtual reality feels off. Adding subtle monster sounds can provide direction and distance clues. Integrating a particle effect such as gray fog can enhance the spooky, mysterious atmosphere. By stimulating multiple senses and adapting these elements to a user's movements, we can significantly enhance the overall immersion and user experience of the AR monster hunt game.

This chapter aims to enhance your XR design skills, targeting more user senses for natural, immersive experiences. We'll delve into sound theory and particle behavior basics, making these physics concepts accessible to all. We will delve into Unity's audio and particle systems to understand how they can help us recreate real-world phenomena. In line with our hands-on approach, you'll apply these concepts in a practical VR project in Unity, creating your most immersive VR experience yet.

Let's dive into this thrilling exploration of creating realistic XR experiences! This chapter guides you on your way through these four sections:

- Understanding sound theory and Unity's audio system
- Preparing the VR drum scene and adding sound effects
- Understanding particle behavior and Unity's particle system
- Adding particles to VR scenes with varying properties

Technical requirements

To fully engage with and benefit from the VR application development process outlined in this chapter, your hardware must meet certain technical specifications. If you followed the tutorial in *Chapter 3* and your Unity setup remains the same, you can bypass this section.

To follow along with this book's content and examples, ensure your computer system can accommodate *Unity 2021.3 LTS* or a more recent edition, with *Android* or *Windows/Mac/Linux Build Support*, depending on the nature of your VR headset and PC.

Understanding sound theory and Unity's audio system

In the subsequent sections, we will delve into the fascinating physics of sound waves and explore how Unity's robust audio system allows us to manipulate these properties with its diverse range of audio components. Our journey begins with an introduction to the fundamentals of sound, the nature of sound itself, and its defining features – frequency and amplitude.

What are frequency and amplitude?

Sound, in essence, is a form of mechanical energy that propagates as a wave through a medium – most commonly, air. Simply put, mechanical energy is energy associated with the motion of objects. When an object vibrates, it displaces the surrounding air particles, setting off a chain reaction of particle displacement that travels outward in the form of a sound wave.

The two main parameters describing a sound wave are frequency and amplitude.

Frequency is the number of complete wave cycles per second, measured in **Hertz (Hz)**. It determines the pitch of the sound – a higher frequency corresponds to a higher pitch, and a lower frequency corresponds to a lower pitch. In game development, understanding frequency is crucial. For instance, a small, light creature such as a bird would have a high-pitched (high-frequency) chirp, while a large creature such as an elephant would have a low-pitched (low-frequency) roar.

Amplitude refers to the maximum displacement of particles by the sound wave. In simpler terms, it's how far the particles are pushed when the wave passes through. It corresponds to the loudness of the sound – a larger amplitude means a louder sound, and a smaller amplitude means a quieter sound. In games and XR applications, amplitude can be used to create a sense of distance. A sound source far from the player would have a smaller amplitude and thus sound quieter, while a source close to the player would sound louder.

Understanding other properties of sound

While frequency and amplitude are fundamental, several other properties can add depth to the sound experience in XR. These include timbre (pronounced *tamber*), envelope, spatial audio, reverberation, and echo. Let's go through them step by step:

- **Timbre:** This describes the color or quality of a sound and allows us to distinguish between different sound sources that may have the same pitch and loudness. Here, *color* refers to the unique characteristics and tonal nuances of a sound that differentiate it from other sounds. It's what allows us to tell the difference between a violin and a flute, even if they're playing the same note at the same volume. In game development, timbre can be used to give different characters or environments their unique sound signatures.

- **Envelope:** The envelope of a sound refers to how it evolves over time. It's traditionally broken down into four parts – **Attack, Decay, Sustain, and Release (ADSR)**. ADSR describes the initial spike of a sound (attack), the subsequent decrease to a stable level (decay), the maintenance of that level for a duration (sustain), and the eventual fading away of the sound (release). In XR, modifying the ADSR envelope can make sounds seem more realistic or can be used creatively to give sounds a stylized feel.

- **Spatial audio:** This refers to the perception of the direction and distance of a sound source. Our brains use several cues to pinpoint the location of a sound source, such as the time difference between the sound reaching our two ears (binaural cues), the change in frequency caused by the shape of our ears (spectral cues), and the change in sound as we move our heads (dynamic cues). Game and XR developers can simulate these effects to create immersive 3D soundscapes, where players can locate a sound source within the game environment. For instance, in a VR horror game, you could use spatial audio to make it sound like eerie whispers are coming from just over the player's shoulder, or like the ominous footsteps of a monster are getting closer and closer. These techniques can significantly enhance immersion and even gameplay, as players could use sound to navigate their environment or detect threats.

- **Reverberation** or **reverb:** This is the persistence of sound in a particular space after the original sound is removed. It's caused by sound waves reflecting off surfaces in the environment, such as walls and floors, creating a multitude of echoes that gradually fade out. This can give players a sense of the size and material of the space they're in – a large, stone-walled cathedral would have a long, bright reverb, while a small, carpeted room would have a short, dull reverb.

- **Echo:** This is the distinct, delayed reflection of sound that can be clearly identified as a repetition of the original source. It's caused when sound waves bounce off distant obstacles and return to the listener after a noticeable time gap. The time it takes for the reflection to return can give listeners an impression of the distance to the reflecting surface – a mountain range would produce a long-delayed echo, while a closer building might yield a quicker, more immediate echo. Different surfaces can also affect the tonal quality of the echo, with hard surfaces such as concrete producing clearer echoes compared to softer surfaces such as dense foliage.

Understanding and manipulating these aspects of sound can allow you to create rich, immersive, and interactive soundscapes in your XR applications. With careful design, sound can be not only an accompaniment to the visual experience but also a key element of gameplay and immersion. The following section introduces you to Unity's audio system, which we will use throughout this chapter to breathe even more life and immersion into our XR experiences.

Exploring Unity's audio system

Sound design is an essential part of creating immersive XR experiences. With Unity's audio system, you can create dynamic, spatialized audio that reacts to the player's actions and movements, enhancing immersion and even guiding gameplay.

For instance, in a VR horror game, you could use spatialized sound to create an ominous atmosphere and build tension. Distant, eerie noises could hint at unseen dangers, while sudden, nearby sounds could provide jump scares. As the player navigates the environment, the sounds change based on their position and orientation, making the virtual world feel alive and reactive.

Understanding and effectively using Unity's audio system is a crucial skill for any XR developer. Sound isn't just an accessory to the visuals; it's also a powerful tool for storytelling, player guidance, and world-building.

Let's now take a more detailed look at the three core components of Unity's audio system – AudioClips, AudioSources, and AudioListeners.

- **AudioClip**: `AudioClip` represents a sound file that can be played back in your application. It holds the data for the sound, but on its own, it can't actually produce sound. Think of an `AudioClip` file like a CD – it holds music, but you need a CD player to actually hear the music. `AudioClip` files are versatile – you can use them for short sound effects, such as a character's footsteps, or for longer pieces of audio, such as background music or dialogue. Unity supports a range of audio file formats, including `.wav`, `.mp3`, and `.ogg`. When designing sound for XR, remember that audio quality is important for immersion. High-quality `AudioClip` files can make a virtual environment feel more real.

- **AudioSource**: `AudioSource` is like a CD player for your `AudioClip` file. It's a component that can be attached to a GameObject to play sounds. You can think of it as a point in your 3D space where sound originates. Every sound you hear in a Unity application originates from an `AudioSource` component. An `AudioSource` component plays back an `AudioClip` file in the scene and controls how that sound is played. The `AudioSource` component offers several properties that you can manipulate:

  - **Clip**: This refers to the `AudioClip` file that will be played.

  - **Play On Awake**: If this is enabled, the `AudioSource` component will start playing its `AudioClip` file as soon as the scene starts.

- **Loop**: If this is enabled, the `AudioSource` component will loop its `AudioClip` file, starting it again as soon as it ends.

- **Volume**: This controls the loudness of the `AudioSource` component, and can be used to fade sounds in or out.

- **Pitch**: This controls the pitch of `AudioSource`. Higher values result in a higher pitch, and lower values result in a lower pitch. This can be used to create a **Doppler effect**, which is a change in the frequency or wavelength of a wave for an observer moving relative to its source – for instance, the sound of a passing car changing from high to low pitch.

- **3D Sound Settings**: These settings control how the sound is affected by distance. You can make the sound get quieter with distance, change the pitch with distance, and so on. This is critical for XR applications, as it helps to create a sense of space and realism.

- **AudioListener**: This is the scene component that hears the sounds. You can think of it as the user's ears in the 3D world. It receives input from all `AudioSource` components in the scene and processes them to create the final mixed sound that the player will hear. In most games and XR applications, the `AudioListener` is attached to the main camera or the player's avatar. As the player moves through the environment, different sounds will get louder or quieter based on their distance and direction from the `AudioListener`, creating a dynamic soundscape.

> **Important note**
>
> It's important to note that you should generally only have one `AudioListener` in your scene. Having more than one can cause sounds to be duplicated, which can result in distortion and other audio artifacts.

The following sections guide you on how to implement these components into your Unity scene through a new, hands-on VR project.

Preparing the VR drum scene and adding sound effects

After exploring the physical properties of sound waves and Unity's audio system, you will finally put this theoretical knowledge into use by building your own VR drum scene.

Once you've created a VR scene featuring various drums, which can be struck by a player equipped with VR drumsticks in each hand, you'll discover how to augment your setup by assigning distinct sound files to each drum that gets hit. Moreover, you'll learn how to adjust the sound volume based on the intensity of the strike delivered by the VR drumsticks, adding an extra layer of realism. This refined VR drumming environment will plunge users into an impressively accurate and engaging drumming experience. Let us now set up and prepare our VR drum scene.

Setting up and preparing your VR drum scene

The following steps guide you on how to create your project, set up the needed VR settings in the Unity Editor, and add a player with a ground floor:

1. Create a new project with the URP by selecting **New Project** in the Unity Hub, choosing the **3D (URP)** template, and renaming the project to a name of your choice, such as Drum Scene.

2. Once the scene has loaded, navigate to **Edit | Project Settings | XR Plugin Management** and click the **Install** button. After the installation, enable the **OpenXR** checkbox. If you are prompted to restart the Editor, follow the instructions to do so.

3. Under **Edit | Project Settings | XR Plugin Management**, go to the **OpenXR** submenu. Under the **Windows/Linux/Mac** tab, select the interaction profile that matches your headset.

4. To add the required prefabs for our VR player to our scene, we need to install the XR Interaction Toolkit. To do so, go to **Window | Package Manager** and click on the + icon in the top-left corner. Choose **Add package by name**, type in com.unity.xr.interaction.toolkit and 2.5.1, and then click the **Add** button. This will automatically install the XR Interaction Toolkit into your scene. Once the installation process is finished, select **XR Interaction Toolkit** in the **Package Manager** window, and click the **Import** buttons next to **Starter Assets** and **XR Device Simulator**, as shown in *Figure 7.1*.

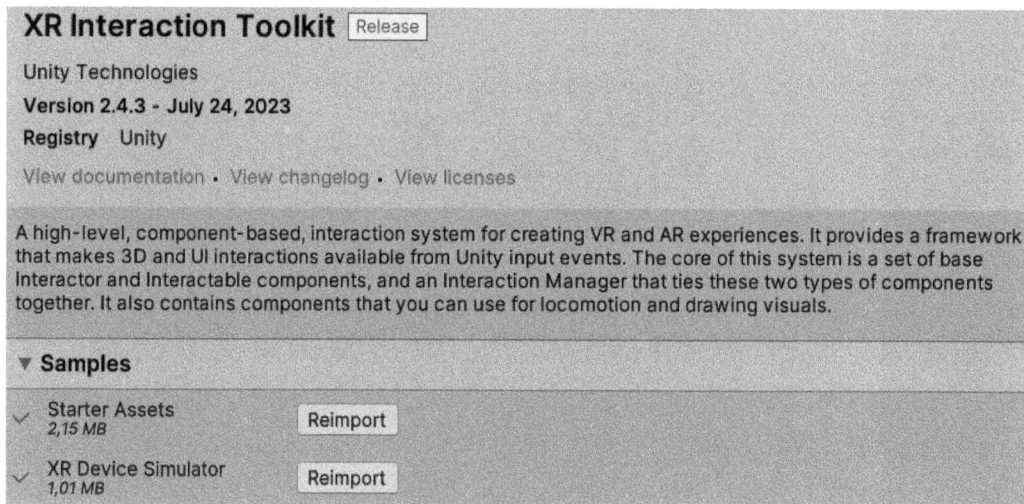

Figure 7.1 – The XR Interaction Toolkit successfully installed alongside the imported
Starter Assets and XR Device Simulator in the Package Manager window

5. Delete the existing main camera in the hierarchy, as the VR player we will add comes with a point-of-view camera. Now, navigate to the Assets folder in the **Project** window and search for the **XR Interaction Setup** prefab. Drag this prefab into your hierarchy. It includes a complete

player setup, allowing movement, teleportation, interaction, and snap-turning. To create a floor for the player, right-click in the hierarchy and select **3D Object | Plane**, rename the plane to `floor`, and place it at the origin (0, 0, 0).

Hurray! You've set up the framework for a VR-enabled drum scene in Unity! The subsequent section will guide you through the process of adding the drums and drumsticks to your scene.

Creating and importing 3D models for your VR drum scene

Although Unity Asset Store provides an extensive collection of different 3D models, it doesn't offer free drum models that are suitable for our project. Therefore, we'll utilize 3D models available on *Sketchfab*, another rich platform that allows you to download free and paid 3D models from a wide range of creators.

The selected drum models for this chapter were found at `https://skfb.ly/o8QvS` and were created by Bora Özakaltun. Open the link and download the free models in the `.fbx` file format, as shown in *Figure 7.2*.

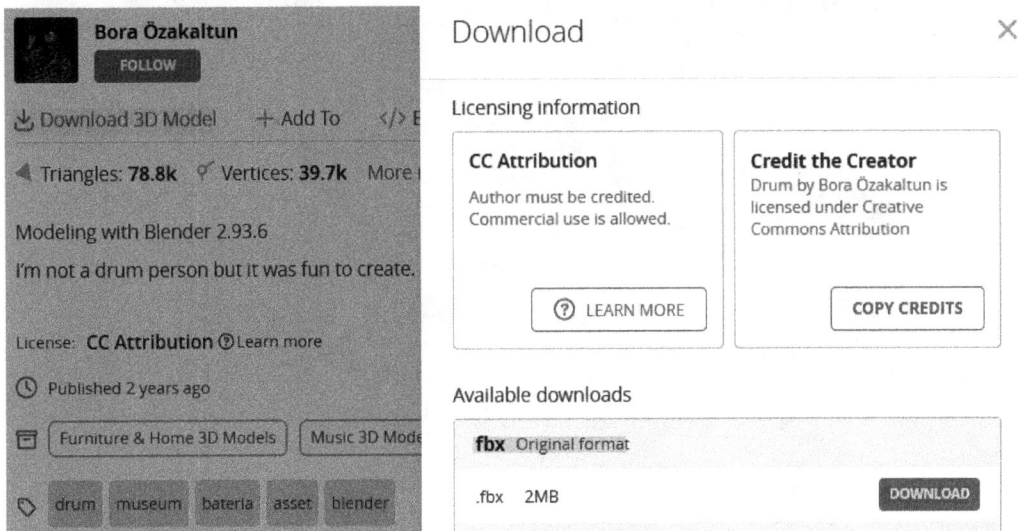

Figure 7.2 – The required mouse clicks (highlighted in yellow) to download the drum models as an .fbx file

Important note

Should the models be unavailable at your time of reading, you can clone the entire project from the GitHub repository for this book and extract the assets from there.

Once the folder is downloaded, unzip all the child folders. Then, drag the unzipped drum folder into your **Project** window, completing the model importation. Now, it's time to integrate the drums into your virtual environment. Follow these steps to position the drums in your scene:

1. In the **Project** window, navigate to the Assets folder, and then to drum | source directory. Here, you will find the drum model that you previously imported.

2. Select the **drum** file and drag it into your scene. Upon doing so, you might observe that the drum appears disproportionately large for the environment.

3. To rectify the size issue, select the drum in the hierarchy and scale it down to the dimensions (0.02, 0.02, 0.02). Place the drum at the origin coordinates (0, 0, 0), aligning it perfectly within the virtual space.

Once these adjustments are complete, your scene should resemble the depiction in *Figure 7.3*.

Figure 7.3 – How your drum scene should currently look in the Game view of the Unity Editor

Now, we need drumsticks to interact with the drums. By following these steps, we will create our own drumsticks within Unity:

1. Right-click in the Scene Hierarchy, navigate to **GameObject | Cylinder**, and scale it to the dimensions (0.01, 0.2, 0.01). Rename this GameObject Drum Stick.

2. As you can see in the **Inspector** window, the drumstick automatically includes a **Capsule Collider** component. A **Capsule Collider** is a 3D shape used in Unity that resembles a capsule or pill-like shape, used to detect collisions between objects. Additionally, it is essential to add a **Rigidbody** component to the drumstick. This allows the object to be affected by forces and physics within the Unity environment, enabling dynamic interactions such as the ones needed for an immersive drumming experience. With the drumstick selected, head over to its **Inspector** window and select the **Add Component** button. Now, add a **Rigidbody** component by searching

for it and selecting it. This component allows Unity's physics engine to treat the drumstick as a physical object, simulating interactions and reactions to forces, which is critical to realistically model the drumstick's behavior when striking a drum.

3. Set **Collision Detection** of the **Rigidbody** component to **Continuous Speculative**. This mode enables more accurate and efficient handling of collisions between the drumstick and other objects, particularly when dealing with fast motions typical of drumming.

4. Create a new material in the `Assets` folder, and name it `Drum Stick Material`. Download `Balsa_Wood_Texture.jpg` from `https://commons.wikimedia.org/wiki/File:Balsa_Wood_Texture.jpg` and import it into the project. Select `Drum Stick Material`, go to **Inspector**, and drag `Balsa_Wood_Texture.jpg` into the **Base Map** field. Apply this material to the drumstick.

5. Select `Drum Stick`, and click on the **Tag** dropdown in the Inspector. Select **Add Tag**, click on the plus icon, type in `Drum`, and save the tag. We will need this tag to address the drumsticks in our scene later on.

6. In the **Project** window, create a folder named `Prefabs`, and drag the drumstick into it. This will create a prefab, allowing easy duplication and modification.

7. Finally, attach the prefab as a child to both the left controller and the right controller, as detailed in *Figure 7.4*.

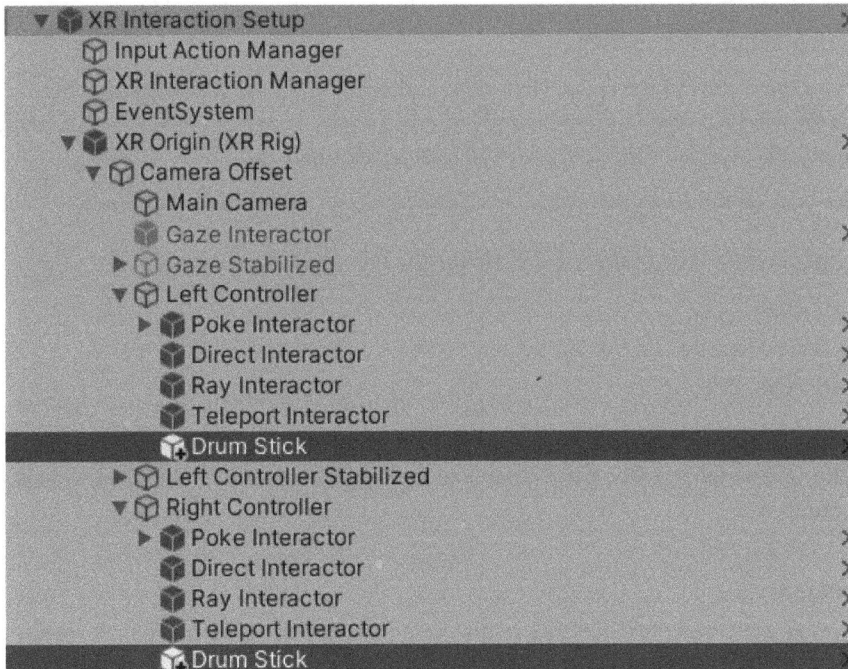

Figure 7.4 – Each drumstick being successfully attached to the respective controller of the XR Origin in the Scene Hierarchy window

You have successfully completed all the steps to set up and prepare your VR drum scene. In the following section, you will learn how you can utilize Unity's audio system to add sound effects to your scene.

Adding sound effects to your VR drum scene

This section guides you through the process of adding sound effects to your VR drum scene. The first step involves triggering audio when hitting an instrument, adding realism and engagement to the scene.

Scripting sound playback on collision for your VR drum scene

To trigger a sound playback once our drumsticks collide with the drums, we must use C# scripting. Follow these steps to correctly prepare your drums for it:

1. In the Unity Editor, press *Ctrl / Cmd* on your keyboard, and *left-click* to select all the drums in your Scene Hierarchy.

2. With the drums selected, add a **Box Collider** component via the **Inspector** window by clicking the **Add Component** button and searching for and selecting the **Box Collider** component.

3. Then, we must ensure that the collider functions as a trigger, allowing it to initiate actions rather than physical interactions. To do this, select the **Is Trigger** checkbox in the **Box Collider** component of each drum in the **Inspector** window. This option transforms the collider into a non-physical boundary that can detect when objects pass through, making it instrumental in triggering sound upon collision.

4. Select all the drums in the Scene Hierarchy, and click on the **Add Component** button in the **Inspector** window. Type in `PlaySoundOnCollision`, and click on **New Script**. This action automatically associates the script with all selected drums.

5. *Double-click* on the `PlaySoundOnCollision` script in the **Inspector** window to open it.

Once the script opens in your preferred IDE, define the following three variables:

```
public AudioClip soundClip;
private AudioSource _soundSource;
public string tag;
```

We will use the public `AudioClip` variable to play the audio, the private `AudioSource` variable to play the clip, and assign a public `tag` variable to the drumsticks so that only these GameObjects trigger the sound.

> **Important note**
>
> In Unity, **tags** are simple text identifiers that can be assigned to GameObjects within the **Inspector** window in the Unity Editor to identify and categorize them. In the following code snippet, you will learn how we can use the `tag` variable to trigger sound.

In the `Start()` function of the `PlaySoundOnCollision` script, add the following line of code to access `AudioSource` at the beginning of the experience:

```
_soundSource = GetComponent<AudioSource>();
```

Since we are not continuously checking for changes, we don't need an `Update()` function in this script. Instead, we will implement an `OnTriggerEnter()` method, which is called when something collides with the trigger collider:

```
private void OnTriggerEnter(Collider other) {
    if (other.CompareTag(tag))
    {
        _soundSource.PlayOneShot(soundClip);
    }
}
```

`OnTriggerEnter()` is a special Unity method that is automatically called by the Unity engine itself when a collision with a trigger occurs. This is also why we don't need to call this function within our script and why we selected the **Is Trigger** checkbox earlier in this setup.

Within the `OnTriggerEnter()` method, we call `CompareTag()`, a built-in Unity function from the `GameObject` class. By calling it inside of an `if` statement, we compare the tag of the colliding object to a string. The use of a tag helps ensure that the sound is played only when the drumsticks collide with the drum, and not when other objects might collide with it. By assigning a unique tag to the drumsticks, similar to the `Drum` tag we used, we can easily and efficiently identify them within the `OnTriggerEnter()` method and play the sound only in response to their collisions with the drum.

> **Important note**
>
> At this point, you might be wondering why we are using `CompareTag()` instead of directly comparing the tags with the equality operator (`==`). The reason for this is that using `CompareTag()` is more computationally efficient. Tags in Unity are stored in an internal hashed format, and `CompareTag()` compares these hashes, while the equality operator would first convert the hash to a string, making the comparison slower.

If the tag comparison returns `true`, the next line is executed. The `_soundSource` variable refers to the `AudioSource` component that we will attach to the drums later on. The `PlayOneShot()` method is another Unity function from the `AudioSource` class. It is used to play an audio clip once without interrupting any sounds currently playing on the same `AudioSource`. This functionality is essential for the authenticity of our drum scene. Instead of hearing each drum hit in isolation, we seek to create a continuous flow of sound. The ability to layer individual hits allows for a more immersive and realistic experience, reflecting the natural overlap that occurs when drums are played in quick succession. The use of the `PlayOneShot()` method ensures that the sounds meld together harmoniously, capturing the essence of a live drumming performance.

By combining these elements, the `OnTriggerEnter()` method ensures that when the drumstick comes into contact with the drum, the specified sound is played, creating an immediate and realistic response within the virtual environment. It's a powerful way to add immersion and interactivity to your VR experience. After finalizing the `PlaySoundOnCollision` script, it is finally time to add sound files for each drum to your scene.

Adding sound files to your VR drum scene

In this section, you will learn how to import sound files and assign `AudioSource` components for each drum. Let's go through this process step by step:

1. Download the appropriate sound files for each of the drums in your Unity scene. You can find royalty-free `.mp3` sound files on websites such as `https://www.freesoundslibrary.com/` or `https://pixabay.com/sound-effects/`. Alternatively, you have the option to utilize the sound files provided within the Unity project linked with this chapter. These files are readily available in this book's GitHub repository.

2. Once you have the necessary sound files on your local system, go to `Assets | Import New Assets` in the Unity Editor. Navigate to the location of your sound files and select the ones you want to import. Click the **Import** button, and the sound files will now be available in your Unity project.

3. Select all drums in the **Hierarchy** window (*Ctrl* / *Cmd* + *left-click*). In the **Inspector** window, click the **Add Component** button, search for the **Audio Source** component, and select it.

4. Now, select each drum individually in the Scene Hierarchy, and navigate to their `PlaySoundOnCollision` script in the **Inspector** window. Click on the small circle next to the **Sound Clip** field that we defined earlier, and choose an appropriate sound file for each drum.

Upon testing, you'll find that the drum scene is already quite immersive. However, a few adjustments can heighten its realism. The following section guides you on how to fine-tune your drum scene.

Fine-tuning the VR drum scene to heighten its realism

The current setup plays the sound at a constant volume, regardless of how hard or soft the drum is hit. To enhance realism, we need to adjust the volume in relation to the collision velocity. To achieve this, we can utilize an existing script from Valve called *VelocityEstimator*. In case you aren't familiar, Valve is the prominent video game developer and distributor behind the Steam gaming platform, and they offer a SteamVR plugin for Unity, along with interesting VR scripts available on GitHub. The `VelocityEstimator` script is available in the GitHub repository of SteamVR's Unity Plugin at this link: `https://github.com/ValveSoftware/steamvr_unity_plugin/blob/9442d7d7d447e07aa21c64746633dcb5977bdd1e/Assets/SteamVR/InteractionSystem/Core/Scripts/VelocityEstimator.cs#L13`.

> **Tip**
>
> When dealing with complex physical calculations (such as the relationship between volume and collision), searching the internet for existing solutions or scripts can save time and effort. In XR development, understanding every aspect of physics is not always necessary, but knowing how to implement physical calculations accurately is crucial.

The purpose of Valve's `VelocityEstimator` script is to calculate and estimate the speed and direction of the GameObject we attach it to – in this case, the drumsticks. When applied to our VR drum scene, this script will facilitate the adjustment of the sound volume based on the drumstick's striking speed, thereby imitating the natural correlation between the force of a drum hit and the resulting sound volume. To add the `VelocityEstimator` script to our drumsticks, let's perform the following steps:

1. Download the `VelocityEstimator` script from GitHub (by selecting *Cmd / Ctrl + Shift + S*).

2. In the **Project** window of the Unity Editor, open the `Scripts` folder, and drag and drop your downloaded file from your local file manager into it.

3. Now, add the `VelocityEstimator` script to both drumsticks as a component by clicking the **Add Component** button and searching and selecting the script. Be sure to enable the **Estimate On Awake** checkbox, as this will facilitate the calculation of the drumsticks' speed. *Figure 7.5* shows how the drumsticks' fully configured `VelocityEstimator` script component should look in the **Inspector** window.

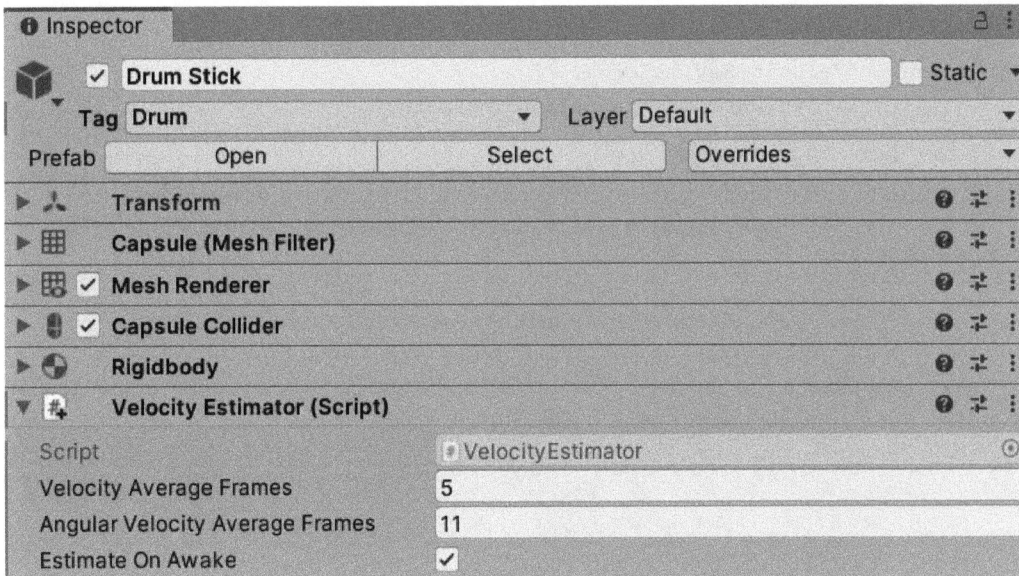

Figure 7.5 – How the drumsticks' fully configured VelocityEstimator
script component should look in the Inspector window

To take advantage of the drumsticks' speed in our existing `PlaySoundOnCollision` script, we must modify it a bit. Open the script again in an IDE, such as Visual Studio, and add the following three lines of code to your script:

```
public bool enableVelocity = true;
public float minimumVelocity = 0;
public float maximumVelocity = 3;
```

The first variable allows us to decide whether we want to take velocity into account when playing the sound. By setting this to `true`, we enable the feature. The `minimumVelocity` variable defines the lower threshold for velocity, allowing us to specify the minimum speed that will impact the volume. Any velocity below this value won't lead to a decrease in volume. Conversely, the `maximumVelocity` parameter sets the upper limit of velocity that will influence the volume. Speeds above this threshold won't result in further increases in volume.

These new parameters grant us control over how the velocity of the drumsticks influences the volume of the sounds produced. By adjusting the minimum and maximum velocity values, we can fine-tune the responsiveness of the drum sounds to create a nuanced and realistic drumming simulation.

Now, we must modify the `OnTriggerEnter()` method to use the `VelocityEstimator` script component if the `enableVelocity` variable is set to `true` like this:

```
    private void OnTriggerEnter(Collider other) {
        if (other.CompareTag(tag))
        {
            VelocityEstimator velEstimator = other.
GetComponent<VelocityEstimator>();
            if (velEstimator && enableVelocity)
            {
                float v = velEstimator.GetVelocityEstimate().
magnitude;
                float soundVolume = Mathf.InverseLerp(minimumVelocity,
maximumVelocity, v);

                _soundSource.PlayOneShot(soundClip, soundVolume);
            }
            else
                _soundSource.PlayOneShot(soundClip);


        }
    }
```

Let's go through the newly added parts of the OnTriggerEnter() method. By calling other.GetComponent<VelocityEstimator>(), we try to get a component of the VelocityEstimator type from the object with the Drum tag. If the VelocityEstimator component is found and the enableVelocity variable is set to true, the code inside the if statement will be executed. In this case, the code first calls the GetVelocityEstimate method from the VelocityEstimator component to get the velocity estimate, and then it takes the magnitude of that vector to get the speed as a single float value. Then, the volume of the sound based on the speed is calculated and stored in the soundVolume variable. The InverseLerp() method returns a value between 0 and 1, representing where the value of v falls between minimumVelocity and maximumVelocity. By calling _soundSource.PlayOneShot(soundClip, soundVolume) on the next line, a one-time sound is played at the calculated volume.

The else statement at the end of OnTriggerEnter() is executed if the previous if statement is not met. The if statement is not met if the VelocityEstimator script component is not found, or if the enableVelocity variable is set to false. The code inside this block will play the sound at its default volume because the velocity of the object was not considered.

If we run the scene right now, we will not hear anything at all when we hit the drumsticks against the drums. This is because, in the original VelocityEstimator script from Valve, the velocity estimation routine was meant to start by calling BeginEstimatingVelocity(). However, in our PlaySoundOnCollision script, this function wasn't being called at all; hence, no velocity estimation took place. This is why we consistently get a velocity of zero right now when testing out the scene and do not hear anything.

To solve this problem, we need to ensure that BeginEstimatingVelocity() is called when the script starts. This can be done by adding the following code lines to the Start() method of our VelocityEstimator script:

```
private VelocityEstimator velocityEstimator;
{

    private void Start()
    {
        velocityEstimator = GetComponent<VelocityEstimator>();
        if (velocityEstimator != null)
        {
            velocityEstimator.BeginEstimatingVelocity();
        }
    }
}
```

By putting the call to BeginEstimatingVelocity() inside the Start() method, we ensure that the velocity estimation begins as soon as the drumstick object is ready, which is exactly what we want.

We've successfully executed all necessary steps to add a sound to each drum that reflects the intensity of each hit. Now, it's time to put on your VR headset and put the final scene to the test. Pay close attention to the variation in volume for each drum sound, depending on the collision velocity with the drumstick. Also, keenly observe the sound dynamics when you strike two drums simultaneously or hit several drums in rapid succession. You'll be amazed by how incredibly realistic our VR drum scene has become.

The next sections will teach you yet another valuable skill to make your scenes more natural and immersive – adding particles!

Understanding particle behavior and Unity's particle system

When designing immersive XR experiences, understanding the physics of particle behavior plays an important role. By leveraging Unity's particle system and the fundamentals of particle physics, you can create rich, dynamic, and realistic effects that enhance the immersion of your virtual environments. In this section, you will learn everything you need to know about real-world particle behavior.

Understanding the behavior of particles in the real world

Particle behavior in the physical world refers to the ways in which small fragments or quantities of matter move and interact, based on forces, environmental conditions, and intrinsic properties. Particles in the natural environment behave according to certain laws and principles. Key factors include gravity, air resistance, life span, and collision behavior. Here is an overview of what all these terminologies mean:

- **Gravitational forces**: Particles are affected by gravitational forces, which pull them toward the center of mass. However, it's important to note that not all particles are affected by gravity in the same way. Consider two common particle systems – rain falling from the sky, and sparks rising from a fire. In the case of rain, gravity pulls the raindrops down toward the ground. Conversely, sparks from a fire rise upward because the heat decreases their effective gravitational pull (hot air rises).

- **Air resistance**: Air resistance, also known as **drag**, is the resistance a particle experiences when moving through a medium such as air. It affects both the speed and the direction of particles, often resulting in less linear, more natural-looking motion. Smoke from a fire or a chimney provides a good example. While the heat and updraft may initially propel the smoke upward, air resistance and wind can cause it to billow, curve, and sway. Similarly, consider a particle system representing leaves blowing in the wind. Air resistance causes the leaves to flutter and spin, rather than moving directly in the direction of the wind.

- **Life span**: Every particle has a life span, a duration of existence after which it disappears or changes state. This life span, along with changes that happen during it, contributes to the believability of a particle effect. Consider a firefly effect, where each firefly would be a particle that appears, glows brightly for a few seconds (reaching peak brightness midway through its life), and then fades away. A snowflake particle system offers another example. As snowflakes, which are represented by particles in your Unity scene, fall toward the ground, they could fade or shrink to give the illusion of snowflakes melting as they touch the warmer ground.

- **Collision behavior**: When particles come into contact with a surface or another particle, they react in a manner that depends on their nature and the surface they collide with. This is known as collision behavior. Raindrops, for example, splash and disappear upon hitting a hard surface, creating smaller droplet particles. Conversely, confetti pieces bounce and scatter instead of splashing when they hit a surface.

Incorporating these principles of particle physics into your Unity Particle System will significantly enhance the realism and immersion of your XR experiences. You will learn about Unity's Particle System in the following section.

Exploring Unity's Particle System

Unity's Particle System is a powerful tool for XR developers that adds another level of immersion to the user experience. It is used to create a wide range of special effects such as fire, smoke, sparks, and magic spells, as well as more abstract visual elements. Understanding and utilizing Unity's Particle System effectively can significantly enhance the visual appeal of your application and deepen the sense of presence within the virtual environment.

The following is a detailed exploration of the core components of Unity's Particle System – specifically, the Particle System component and the Particle System Renderer component. Both of these components will be vital later on when we add a particle system to our drum scene:

- The **Particle System component** is the main engine of the particle system in Unity. The Particle System component itself is attached to a GameObject and controls how particles are generated and behave over their lifetime. It offers a multitude of modules, each controlling a different aspect of the particle's behavior:

  - **Emission Module**: This controls the rate at which new particles are spawned. Whether you need a constant trickle of particles or a sudden burst, this module has you covered.

  - **Shape Module**: This defines the region and the form from which particles are birthed. This could be a simple point, a complex mesh, or anything in between, offering a flexible start to your particles' life journey.

- **Gravity Modifier**: This is a setting within the Particle System component that emulates the influence of gravity on particles. You can adjust this setting to make the particles fall faster or slower, allowing you to create effects such as floating dust or rapid rainfall.

- **Velocity over Lifetime Module**: This dictates how a particle's speed and direction change over its lifetime. Coupled with the Gravity Modifier, this can create lifelike effects of particles being caught in wind or turbulence.

- **Drag**: Found under the **Forces over Lifetime** module, which in Unity's Particle System allows users to apply varying forces to particles throughout their life span, this property lets you simulate the effect of air or fluid resistance. By modifying the Drag property, you can make particles move as if they are in a heavier medium, providing a sense of weight and depth to the particles.

- **Color over Lifetime Module**: This specifies how a particle's color evolves over its life span. This module, in conjunction with the Size over Lifetime module, allows you to create a natural fade-in and fade-out effect, enhancing the realism of particles.

- **Size over Lifetime Module**: This determines how a particle's scale changes over its life cycle. By making particles shrink or grow over time, the environment seems to evolve and be dynamic to observers.

- **Collision Module**: This module governs how particles interact with other GameObjects in the scene upon collision. You can control properties such as bounce (restitution), dampen (loss of speed), and lifetime loss on collision. This can offer a high degree of realism to how particles respond to their environment, such as sparks bouncing off a surface or water droplets splashing.

To illustrate the application of these principles, consider the example of a simple campfire. In this case, embers rising from the fire can be created as a particle effect where the particles move upward, affected by a slightly randomized velocity to simulate the effect of heat and air resistance. Gravity's influence would be negative here (pulling particles upward) due to the heat, and particles could have a reddish color at birth, fading to dark as they cool, simulating the life cycle of real embers. The smoke, however, can be created with particles moving upward with a higher randomized velocity, simulating the churning effect of fire. These particles would be less affected by gravity and have a longer life span. They could also use a color gradient, changing from dark gray near the fire to a lighter color as they rise and cool. Lastly, the fire itself can be simulated with a high rate of small, bright particles with short lifetimes and high randomized velocity. The effect would be a vibrant, dynamic flickering fire.

- The **Particle System Renderer component** is responsible for rendering the particles on the screen. This component can be customized to fit the specific visual needs of your application. Some of its properties include the following:

  - **Material**: This defines the appearance of the particles and can include textures, colors, and shaders

- **Render Mode:** This determines how the particles are rendered and can be in the form of billboards (always facing the camera), meshes, or other similar entities

- **Sorting Mode:** This determines the order in which particles are rendered and is especially important when particles overlap

Remember, particles aren't just visual fluff. They can play a crucial role in your storytelling, player guidance, and world-building efforts. For instance, a trail of mystical sparkles might guide a player toward a hidden treasure, or a plume of smoke could hint at a recently extinguished campfire nearby.

Now that we've explored the key components of Unity's particle system, let's dive into the following section, which explains how to implement a particle system into our drum scene. The goal is that every time we hit a drum, fog is released, proportionate to how hard we hit it.

Adding particles to VR scenes with varying properties

Do you recall the last time you attended a live concert when, at the pinnacle of a legendary song, the stage was enveloped by a blanket of white fog while you were lost in the music with your friends? Taking inspiration from the theatrical smoke often utilized in real-world concerts, we're going to apply our recently acquired knowledge about Unity's Particle System to create a similar effect. Our objective is to integrate a particle system into our drum scene to release fog every time a drum is hit, with the intensity of the fog corresponding to the velocity of the strike. The more frequently the drums are played, the more the fog should saturate our scene, and vice versa, thereby fully replicating the euphoric sensation of being at a genuine concert.

To achieve this objective, we need to first initialize a Particle System in our scene.

Initializing a Particle System in your VR drum scene

Follow these steps to establish a Particle System in your VR drum scene:

1. In your Unity scene, *right-click* on the **Hierarchy** window, navigate to **Effects**, and select **Particle System**. This will instantiate a new Particle System within your scene.

2. The existing Particle System employs Unity's standard particle material, which isn't ideal for creating fog. However, we can easily customize our own material for the Particle System. Go to the **Project** window, *right-click*, and select **Create | Folder**. Name this new folder Fog.

3. Download a transparent image of a cloud or fog. We will use a cloud image, which is available for download at this link: `https://pixlok.com/images/clouds-png-image-free-download/`. You can also access it through the GitHub folder for this chapter. Drag and drop this image into the Fog folder you just created.

4. Within the Fog folder, generate a new material by *right-clicking* and then selecting **Create |
    Material**. Name this new material Fog. Now, drag and drop the previously downloaded image
    from your local filesystem into the base map of the material. Next, modify the **Shader** field of
    the material to **Mobile/Particles/Alpha Blended** in the **Inspector** window. This setting allows
    the particles to overlap with each other and blend seamlessly, creating a more realistic fog effect.

5. Select the Particle System, and drag the Fog material into its **Inspector** window. This will
    automatically update the material. As a result, your scene should now resemble the scene
    shown in *Figure 7.6*.

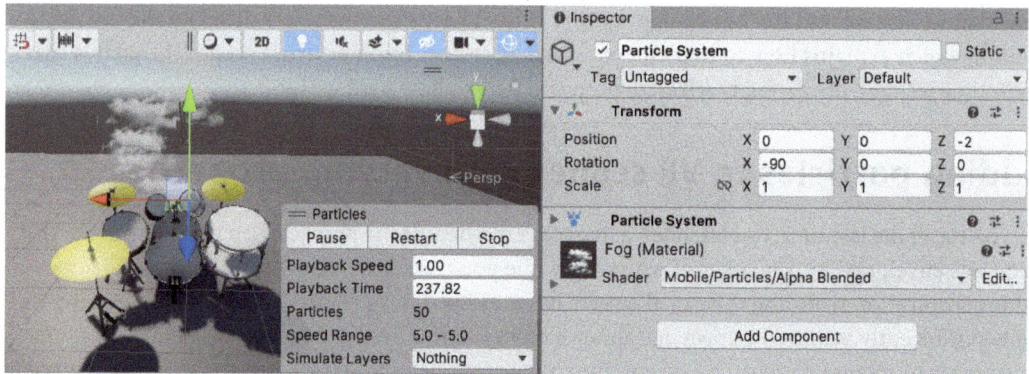

Figure 7.6 – How your current Unity scene and the Inspector window of
the Particle System should look with the newly added fog

After successfully adding the Particle System to our VR drum scene, let's modify its properties a bit
to achieve a more dynamic behavior for the fog, which is dependent on the collision velocity of the
drums and drumsticks in the next section.

Modifying the properties of the Particle System in your VR drum scene

Let's explore the properties of the Particle System in the **Inspector** window by expanding it. There
are a few critical properties that you need to modify to make the fog on your VR drum scene behave
naturally. The modified properties can be observed in *Figure 7.7*.

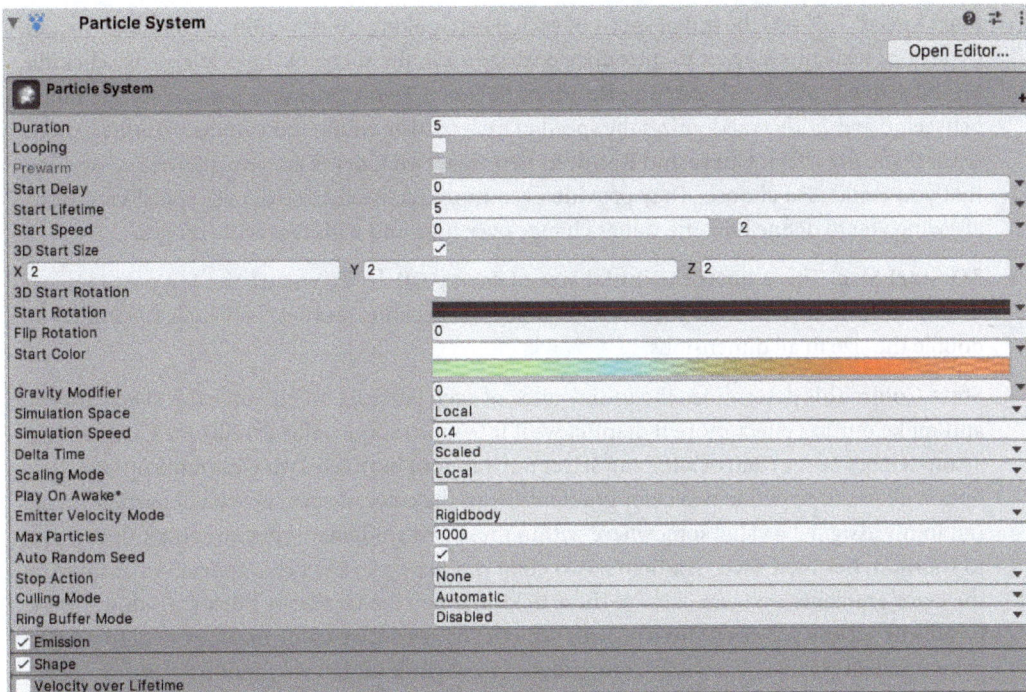

Figure 7.7 – The expanded Particle System in its modified form in the Inspector window

These properties are explained in the following list in detail:

- **Duration**: This determines how long the Particle System will emit each particle. If the **Looping** option is enabled, the system will reset and commence a new cycle after the designated duration is completed. In our case, we will choose a duration of 5 seconds to make the scene more dynamic.

- **Looping** and **Play On Awake**: **Looping** refers to the continuous replay of the particle sequence, while **Play On Awake** ensures the Particle System starts playing as soon as its object is activated. Both options can be disabled in our case, since we'll be triggering particle emissions via the `PlaySoundOnCollision` script.

- **Start Delay**: This introduces a delay before the Particle System begins to emit particles. In our case, we want the particle emission to start immediately upon drum impact, so we will set this value to 0.

- **Start Lifetime**: This dictates the initial life span of each particle, measured in seconds. For our purpose, a 5-second life span should be sufficient.

- **Start Speed**: This sets the initial speed of each particle along the direction of emission. To add a touch of realism, we want to introduce a variation in the speed of the particles. By clicking on the arrow symbol and selecting **Random between Two Constants**, we can set the range between 0 and 2. This randomizes the speed of each particle within these limits, creating a more naturalistic fog effect. **Curve** and **Random between Two Curves** are two alternative options that you could also choose. They provide even more nuanced control over speed variation, allowing you to define how the values change over time and within specific ranges.

- **3D Start Size**: This adjusts the initial size of each particle. To ensure the fog particles are discernible, we will increase their size. Enable the checkbox and set the values to (2,2,2) to double the size in all dimensions.

- **Start Color**: This determines the initial color of each particle. To enhance the visual appeal and make our fog effect more dynamic, we'll introduce some color variations. Click on the arrow symbol next to **Start Color** and select the **Random between Two Gradients** option. This option allows us to define two color gradients, with the color of each individual particle being randomly assigned a value somewhere within these two gradients. After switching the option to **Random between Two Gradients**, two color gradient preview fields appear. To customize the color gradients, you can click on these fields to open the **Gradient Editor** window. In the **Gradient Editor** window, you can add, remove, or rearrange color markers to achieve the color gradient you desire. *Figure 7.8* provides a visual guide on how to modify a color gradient.

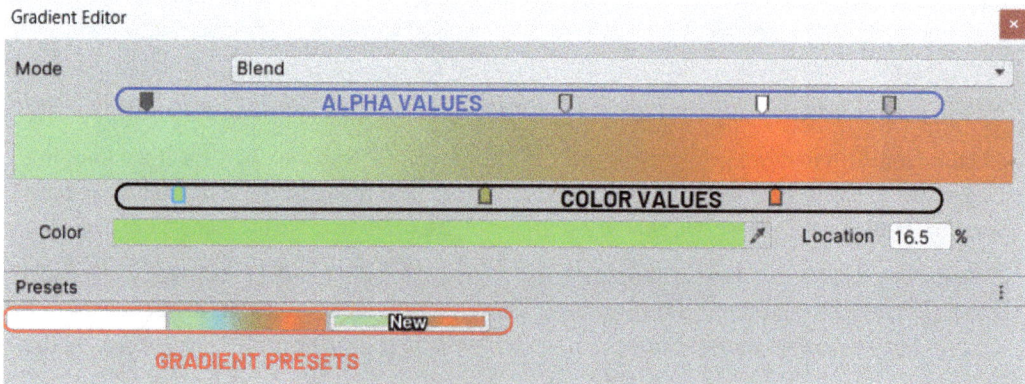

Figure 7.8 – How to modify the Gradient Editor of the start color

Initially, you can set **Mode** to either **Blend** or **Fixed**. The **Fixed** mode presents a solid color, whereas the **Blend** mode enables smooth transitions between colors. You can then add multiple keyframes by simply *left-clicking* either above or below the color space. The upper keyframes establish the alpha values, dictating the transparency, while the lower keyframes determine the colors themselves.

At the bottom of the window, you'll see two existing gradient presets, with a third, labeled **New**, that corresponds to the gradient we're currently editing. Create two gradient presets as per your preference, and assign them to the start color. In our case, we will choose to use the two pre-existing presets that we had (the white and rainbow gradients). At runtime, this configuration will yield a captivatingly vibrant and dynamic color spectrum in our fog effect, as shown in *Figure 7.9*.

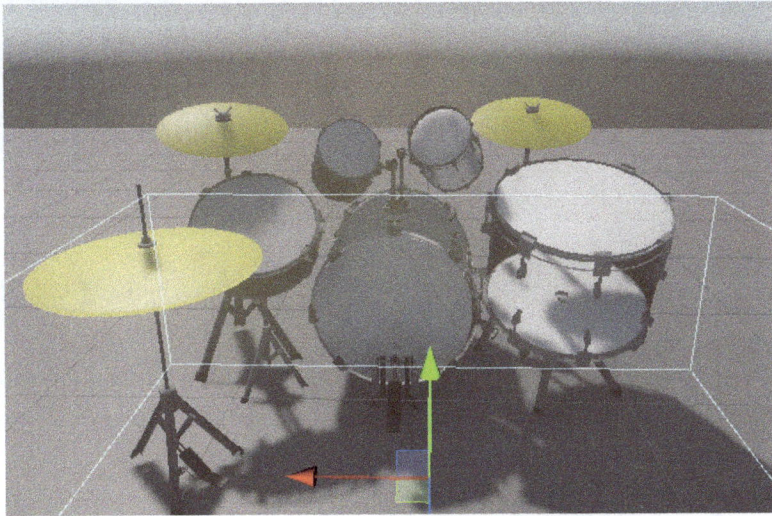

Figure 7.9 – The visual appearance of the fog when choosing two pre-existing presets for the start color

- **Gravity Modifier**: This parameter determines the degree to which each particle is affected by gravity. For our purposes, we don't want gravity to have an impact, so we will set this to 0.

- **Simulation Speed**: This is a multiplier for the Particle System simulation speed. To slow down our fog's movement, we can lower it to 0.4. If the simulation speed is adjusted to a value lower than 0.4, the fog's movement will be even slower, creating a more languid appearance, whereas a value higher than 0.4 will make the fog move more rapidly, giving it a brisker motion.

- Expand the **Emission** module, and adjust **Rate over Time** to 13. **Rate over Time** determines the number of particles emitted per second, and in this context, a value of 13 is chosen as the ideal rate to achieve the desired particle density and appearance for our purpose. By keeping **Rate over Distance** at 0, we ensure that particles are emitted based on time alone, not movement, giving us precise control over the number of particles being emitted.

- Further down, expand the **Shape** module to modify the shape and direction of the particles being emitted. To achieve a fog effect, we need a **Box** shape. We can scale this shape to (3,3,1) to create a larger fog effect. Due to the Particle System's -90-degree rotation on the *X* axis, the *Z* axis now points in the *Y* direction and doesn't need to be scaled.

- Enable the **Color over Lifetime** module, which we learned about in the theoretical section, and set one **Alpha** keyframe in the middle of the color gradient with a low **Alpha** value of 30. This will gradually fade the particles, adding to the realistic fog effect. *Figure 7.10* provides you with a comparison of how the fog looks with and without the **Color over Lifetime** module enabled, and with a low **Alpha** setting.

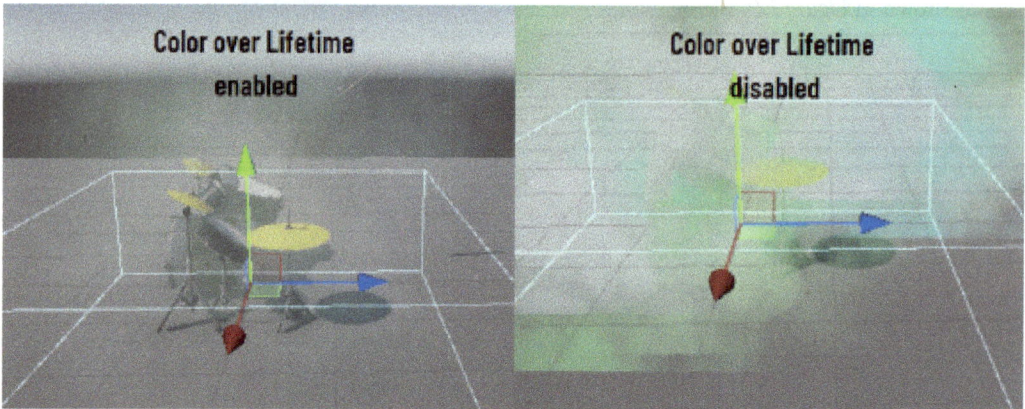

Figure 7.10 – Comparing the visual appearance of the fog with or without the Color over Lifetime module enabled

While the Particle System provides a plethora of parameters to play around with, the ones we've discussed are the most crucial to creating a realistic fog effect. If you're interested in diving deeper into Particle Systems, Unity offers a short, free course that we highly recommend: `https://learn.unity.com/tutorial/introduction-to-particle-systems#`.

Scripting fog appearance on collision for your VR drum scene

After setting up the particle system, we need to go back to the `PlaySoundOnCollision` script and make a few adjustments.

Firstly, we need to create a `public` variable that can reference the Particle System we want the particles to emit from. The code for this is relatively simple:

```
public ParticleSystem fogParticleSystem;
```

Then, whenever the drum gets struck, we should call a method that enables the emission of particles from the Particle System. We base the number of emitted particles on the speed of the drumstick's impact. Since we already have the `OnTriggerEnter()` method that adjusts the sound volume based on the velocity of the drumstick, we can simply apply the same logic to the particle system. To do this, we add three lines of code to the `if` statement in the `OnTriggerEnter()` method:

```
if (velEstimator && enableVelocity)
{
```

```
    float velocityMagnitude = velEstimator.GetVelocityEstimate().
magnitude;
    float soundVolume = Mathf.InverseLerp(minimumVelocity,
maximumVelocity, velocityMagnitude);

    _soundSource.PlayOneShot(soundClip, soundVolume);

    // Convert the velocity to an integer representing the number of
particles
    int numParticles = Mathf.RoundToInt(velocityMagnitude * 10f);

    // Create an EmitParams instance
    ParticleSystem.EmitParams emitParams = new ParticleSystem.
EmitParams();

    // Emit the particles
    fogParticleSystem.Emit(emitParams, numParticles);
}
else
{
    _soundSource.PlayOneShot(soundClip);
}
```

fogParticleSystem is a reference to the Particle System GameObject we set up previously, which will be assigned through the **Inspector** window in the Unity Editor. The int numParticles = Mathf.RoundToInt(velocityMagnitude * 10f); line translates the velocity's magnitude into a corresponding number of particles, multiplying it by 10 – an arbitrary factor that can be adjusted for your specific needs. The number is rounded to the nearest integer, as the particle emission method requires an integer input.

The ParticleSystem.EmitParams emitParams = new ParticleSystem. EmitParams(); line initializes an instance of EmitParams. This struct can be used to change specific parameters of the Particle System when particles are emitted via scripting. In our case, we will use the default settings of the created Particle System.

Finally, fogParticleSystem.Emit(emitParams, numParticles); calls the Emit() method to instantly emit a defined number of particles. It uses the previously created EmitParams instance and the calculated number of particles derived from the object's velocity.

With these adjustments, the desired functionality is added. Back in the Unity Editor, simply drag and drop the Particle System into the fogParticleSystem field in your PlaySoundOnCollision script component of your drums via the **Inspector** window. Once done, run your scene, and test the implementation while wearing your VR headset. Note how the more frequently the drums are hit, the more the fog saturates the scene, and vice versa.

Congratulations! You now know how to fully replicate the euphoric sensation of being at a concert featuring a real-life drummer!

Summary

In this chapter, we embarked on a fascinating journey through the world of sound and particles, understanding their physical properties and delving into the implementation of these real-world phenomena in Unity scenes for heightened immersive experiences in XR.

Upon reaching the end of this chapter, you should now be capable of not only constructing end-to-end XR applications with intricate interactions or animations but also feel comfortable enhancing these creations, by effectively using Unity's audio and Particle Systems to introduce an additional layer of realism to your XR scenes.

The XR development concepts we've covered so far in this book have largely been aimed at those at the beginner to intermediate levels. However, as we move forward, we have something truly unique and enriching in store for you. The following chapter will further elevate your XR development skills by introducing you to some of the most significant and advanced techniques in this field, such as hand-tracking, eye- and head-tracking, and multiplayer functionalities. This crucial knowledge will refine your skillset, transforming you into a more versatile XR developer and enabling you to create an even broader array of sophisticated XR applications.

Part 3 –
Advanced XR Techniques: Hand-Tracking, Gaze-Tracking, and Multiplayer Capabilities

Congratulations on completing your journey from being a beginner in XR development and Unity to becoming highly proficient in creating various XR technologies with sophisticated logic! In this final part of our book, we aim to further enhance your skills to become an intermediate-level XR developer. We will introduce you to advanced XR techniques that are not only cutting-edge in the XR application landscape but will also elevate your XR scenes to a new level of intuitiveness and enjoyment.

This section will also introduce you to the different phases and aspects of the XR development life cycle. It will provide you with a precise, research-based overview of the current state of the art and future trends in XR development. By delving into additional XR toolkits and plugins beyond those covered earlier in this book, we will prepare you with the knowledge needed to dive deeper into your preferred area of XR development once you've completed this book.

This part contains the following chapters:

- *Chapter 8, Building Advanced XR Techniques*
- *Chapter 9, Best Practices and Future Trends in XR Development*

8

Building Advanced XR Techniques

Up until now, you have explored various parts of XR development, crafting immersive and interactive XR applications suited for a wide array of use cases. As you venture further into mastering XR development, it's essential to not only be proficient at creating basic to intermediate XR applications but also master advanced techniques that elevate the commercial viability and influence of your XR offerings.

This chapter is designed to familiarize you with crucial, high-level XR methods, all through a hands-on approach. You will dive into hand-tracking integration, use eye- and head-tracking for complex interactions, and discover how to establish a multiplayer server to create an engaging multiplayer XR experience.

The content of this chapter might sound intimidating but don't worry – we're here to guide you every step of the way as you incorporate these sophisticated XR strategies into various scenes. Leveraging the solid XR foundation you've built from earlier chapters, you'll find these advanced techniques more intuitive than anticipated.

Regardless of the XR equipment you possess, this chapter promises a wealth of knowledge, with all techniques being adaptable to different setups. We'll delve deeper into this in the *Technical requirements* section.

This chapter includes the following sections:

- Adding hand-tracking to XR experiences
- Interacting with objects in XR experiences via eye- or head-tracking
- Building a VR multiplayer application

Technical requirements

Navigating the hand-tracking, eye-tracking, and multiplayer aspects of this chapter requires understanding both shared and unique technical prerequisites. For a seamless development experience, we recommend acquainting yourself with all listed requirements. Though this chapter delves into advanced topics and may initially seem daunting compared to earlier sections, you can be sure that all tutorials in this chapter are designed for straightforward execution – even if you don't have a VR headset at hand.

First, let's address the overarching technical requirements. To follow along with the tutorials in this chapter, you'll need Unity *2021.3 LTS* or a newer version. Validate your hardware's suitability by comparing it to the system requirements described on Unity's website: `https://docs.unity3d.com/Manual/system-requirements.html`. Depending on your VR headset's specifications, ensure that your setup supports *Windows/Linux/Mac* or *Android Build Support*.

Most contemporary VR headsets, particularly those with inside-out camera tracking, incorporate hand-tracking capabilities. Examples include the Meta Quest series, HTC Vive Cosmos Elite, PlayStation VR2, Pico Neo 2 Eye, the Lynx-R1, Valve Index, HP Reverb G2, Varjo VR-3, and the soon-to-be-released Apple Vision Pro. Always refer to the technical specifications of your VR headset to find out whether it supports hand-tracking.

If you don't have access to a headset that supports hand-tracking, you can still follow this chapter as the XR Device Simulator can replicate the hand-tracking features of a VR headset flawlessly.

Eye-tracking in VR is an evolving field, and interestingly, you can delve into the eye-tracking tutorials of this chapter's eye- and head-gaze-tracking section regardless of your VR headset's specifications. Even if you're working solely with the XR Device Simulator without a physical VR headset, you'll find our tutorial accessible. While we won't spoil all the details now, it's worth noting that the XR Interaction Toolkit provides innovative solutions for situations where a VR headset lacks standard eye-tracking capabilities. As of the time of this book's publication, VR headsets that offer eye-tracking include the PlayStation VR2, HP Reverb G2 Omnicept Edition, Pico Neo 3 Pro Eye, HTC Vive Pro Eye, Pico Neo 2 Eye, and Meta Quest Pro. Furthermore, the upcoming Apple Vision Pro is anticipated to feature eye-tracking. This list might not cover all available options by the time you read this book, so always check the specifications of your VR headset to confirm eye-tracking support.

Now that your hardware is all set up, let's start exploring hand-tracking using the XR Interaction Toolkit.

Adding hand-tracking to XR experiences

In this section, you will learn how to add a hand-tracking functionality to your XR experiences. Before creating a new Unity project, however, you must understand the technical concept of hand-tracking and how far it can enrich an XR experience compared to regular controllers.

Understanding hand-tracking and potential use cases

At its core, **hand-tracking** in VR refers to the technological capability of directly detecting, capturing, and interpreting the nuanced movements and positioning of a user's bare hands and fingers within a virtual environment.

Unlike controller-based tracking, which relies on external devices to mediate and translate user inputs into VR actions such as an Xbox controller, hand-tracking operates without intermediary hardware, offering a direct mapping of real-world hand gestures and movements into the virtual realm. This approach leverages sophisticated sensors, cameras, and algorithms to construct a real-time, dynamic model of the user's hand. From grabbing objects to casting spells with finger gestures, hand-tracking provides a vast array of potential interactions. It facilitates complex and subtle interactions that are hard to replicate with traditional controllers, permitting more organic and intuitive interactions within VR.

For a VR headset to harness the potential of hand-tracking, it should be equipped with high-resolution sensors and cameras capable of capturing detailed movements, down to the subtle motions of individual fingers. These cameras often need to be oriented in a manner that provides a wide field of view to consistently track hands as they move.

Hand-tracking requires real-time interpretation of complex hand and finger movements, necessitating robust processing power. The VR headset should have an onboard processor or be connected to a machine that can handle these computations without latency.

Beyond hardware, the headset's software must be designed or adaptable to recognize and interpret hand movements effectively. This includes having algorithms capable of differentiating between intentional gestures and inadvertent hand motions.

Building on these foundational requirements, the next section will guide you through setting up our Unity project to effectively utilize and enable hand-tracking, unlocking a richer, more immersive experience for users of your XR experiences.

Implementing hand-tracking with the XR Interaction Toolkit

To kick off our process of adding hand-tracking capabilities to an XR project, we must perform the following steps:

1. Go to Unity Hub and create a new project by navigating to the **Projects** tab, clicking the **New project** button, selecting the **3D (URP)** template, renaming it to `AdvancedXRTechniques`, and clicking the **Create project** button.

2. In the **Scene Hierarchy** window, you will see that your scene only contains **Main Camera** and **Directional Light**. You can delete **Main Camera** as it is not needed.

3. Import **XR Interaction Toolkit** into the scene by navigating to **Window | Package Manager**, clicking the + button, selecting the **Add package by name** option, typing in `com.unity.xr.interaction.toolkit`, and hitting *Enter*. The toolkit should now be automatically added to your project. Staying in the **Package Manager** window, navigate to the **Samples** tab of the newly added **XR Interaction Toolkit** package and import **Starter Assets**, **XR Device Simulator**, and **Hands Interaction Demo** by clicking the **Import** button next to each of them.

4. To enable hand-tracking in our scene, we don't only need **XR Interaction Toolkit**, but also Unity's **XR Hands** package. Import it into the project by repeating the previous step, typing in `com.unity.xr.hands`, and hitting *Enter*. Once the package has been added to your project, navigate to the **Samples** tab of the **XR Hands** package in the **Package Manager** window and click the **Import** button next to the **HandVisualizer** sample.

5. Now, we would typically drag and drop the **XR Interaction Setup** component into our scene. However, to add hand-tracking capabilities to our VR scene, we need to import a different prefab that is specifically aimed at this purpose and called the **XR Interaction Hands Setup** prefab. To add it to your scene, type `XR Interaction Hands Setup` into the search bar of the **Project** window, select the prefab, and drag it into the **Scene Hierarchy** window. Position it in the center at (0,0,0).

6. It is time to set up **XR Plug-in Management** correctly to enable hand-tracking. Navigate to **Edit | Project Settings | XR Plug-in Management** and select the **OpenXR** checkbox on either the **Windows/Mac/Linux** tab or **Android**, depending on the needs of your VR headset.

7. Once the **OpenXR** plugin has been installed, navigate to its subtab underneath the **XR Plug-in Management** tab on the left-hand side. Here, go to either the **Windows/Mac/Linux** or **Android** tab. Select the + button to add an **Interaction Profile** item to your project. Besides selecting the controllers of your VR headset in the newly opened menu, you should also add another **Interaction Profile** called **Hands Interaction Profile**. At its core, **Hand Interaction Profile** in **OpenXR** provides a standardized way to interpret hand gestures and movements across different VR headsets. Different VR headsets might have their own technology and methods for tracking hands. Without a standardized system, developers would need to write unique code for each headset's hand-tracking system, which can be time-consuming and impractical.

8. Staying in the **OpenXR** subtab, select the **Hand Interaction Poses** and **Hand Tracking Subsystem** checkboxes, regardless of which VR headset you have.

    When you check **Hand Interaction Poses**, you are telling Unity to use a standard set of poses or gestures defined by the **OpenXR** standard. These include grabbing (grip), pointing (aim), pinching, and poking. So, instead of manually coding the detection of these gestures, Unity does it for you based on this standard. By selecting the **Hand Tracking Subsystem** checkbox, Unity uses the **OpenXR** standard to keep track of where the hands are and how they move. This subsystem is like the engine under the hood that keeps an eye on hand movements.

If you have a Meta Quest device, you must also select the **Meta Hand Tracking Aim** checkbox. This feature enhances the existing hand-tracking by also understanding the direction or aim of the hands, giving your VR app a better sense of where users are pointing or what they might be trying to interact with. While our **Hand Interaction Profile** provides a base layer of understanding hand movements, these checkboxes dive deeper into gesture specifics, actual tracking, and device-specific features.

9. Now, let's add a simple plane to the scene functioning as a ground floor and a cube to interact with testing out hand-tracking. You can achieve this by right-clicking in the hierarchy and selecting **3D Object** | **Plane**. Rename the plane to Ground Floor and position it at the origin (0,0,0).

Repeat the previous step but instead of selecting **Plane**, select **Cube**. Rename **Cube** to Hand Tracking Cube. Position it at (0,1.25,1) and scale it to (0.1,0.1,0.1). In the **Inspector** window of Hand Tracking Cube, click the **Add Component** button and type XR Grab Interactable in the search bar. Select the **XR Grab Interactable** script by double-clicking on it. You should see that a **Rigidbody** component has been automatically added to **InteractableCube**, alongside the **XR Grab Interactable** script. Inside the **Rigidbody** component, make sure that the **Use Gravity** and **Is Kinematic** checkboxes are selected so that our interaction with the cube is possible while obeying the laws of gravity.

> **Important note**
>
> If you don't have access to a VR headset with hand-tracking capabilities, you can simply test this feature by using the XR Device Simulator, as detailed in the *Installing the XR Device Simulator* and *Using the XR Device Simulator* sections of *Chapter 3*. To switch to hand-tracking mode, simply press the *H* key.

It's time to test out the scene using your VR headset. Start the scene as you typically would. You don't even need to have the VR headset's controllers on hand. Once the scene is running, adjust your head so that the external cameras on the VR headset are directed at your hands.

You should notice that the controller visuals have been replaced by hand visuals, mirroring the exact position and movements of your real hands, as shown in *Figure 8.1*.

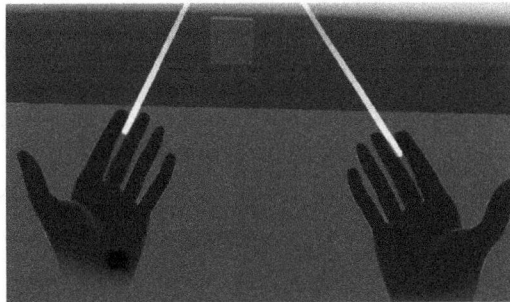

Figure 8.1 – The hand visuals you will see once you put away your
controllers and direct your gaze toward your hands

Try moving each finger separately, shift your hands around, and even hide one hand behind your back. If everything was set up correctly and your VR headset fully supports hand-tracking, the virtual hands should mimic your real hand movements accurately. If you hide one hand behind your back, its corresponding virtual hand should vanish too.

Finally, let's test the interaction with the cube in your scene using only hand-tracking. Aim your right hand's laser pointer at the cube. Touch the tips of your right thumb and right index finger together, forming a near triangle shape, as shown in *Figure 8.2*.

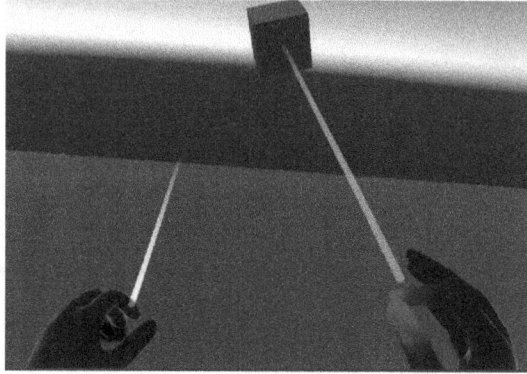

Figure 8.2 – The overlayed hand visuals as you touch your right
thumb and right index finger together in real life

You are now grabbing the cube. While maintaining this position, point in various directions and observe the cube moving correspondingly.

Congratulations! You've now achieved a similar interactive experience in your VR scene with hand-tracking as you would have using standard VR controllers.

To explore how the hand-tracking features of the XR Interaction Toolkit work with other types of objects such as UI elements or buttons, delve into **Hands Interaction Demo**, which is available within the toolkit's **Samples** area that we imported at the beginning. To access it, follow these steps:

1. Head to **Assets** in the **Project** window. From there, navigate to **Samples | XR Interaction Toolkit | Version Number | Hands Interaction Demo | Runtime**.

2. Then, double-click on the **HandsDemoScene** Unity scene. If you prefer, you can quickly locate **HandsDemoScene** using the search bar in the **Project** window.

3. Once you've opened the scene in the Unity Editor, hit the **Play** button. This will let you experience firsthand how to engage with UI elements, press buttons, and even manipulate 3D objects using just your hands.

By now, you should feel like a hand-tracking expert. The next section will introduce you to another advanced concept of XR development: eye-tracking.

Interacting with objects in XR experiences via eye- or head-tracking

In this section, you will not only learn about eye-tracking itself and how it can enrich your XR experience, but you will also implement eye-tracking functionalities to your XR experiences, regardless of whether your VR headset supports eye-tracking or not.

Understanding eye-tracking and potential use cases

From reading a person's intent to enhancing digital interactions, eye-tracking technology is revolutionizing how technologies of all kinds perceive and interpret human behavior.

The eyes are often dubbed the "windows to the soul" due to their ability to express and convey emotions, intent, and attention. Biologically speaking, several key aspects of the eyes play into this:

- **Pupil dilation**: Often a subconscious response, the pupils can dilate or contract based on emotional states, levels of attention, or reactions to stimuli. For instance, someone's pupils might dilate upon seeing someone they're attracted to or contract when exposed to bright light.

- **Saccades**: These are rapid, jerky movements of the eyes when they change focus from one point to another. Often unnoticed by us, saccades play a pivotal role in how we gather visual information from our surroundings.

- **Blinking and micro-expressions**: The rate of blinking can indicate various states, from relaxation to stress. Furthermore, subtle movements around the eyes can give away fleeting emotions – these are known as micro-expressions.

Eye-tracking technology revolves around monitoring and recording the eye's movement and gaze points. Here's how it typically works:

- **Light source**: Infrared light is directed toward the eyes. This light reflects off the cornea and retina.

- **Sensors and cameras**: These detect the reflected light off the eyes. Advanced systems might use multiple cameras from different angles to capture a three-dimensional view of the eye's movement.

- **Data processing**: The raw data captured by the sensors is processed using algorithms to deduce the direction of the gaze and the point of focus.

- **Representation**: The gaze data is usually represented as a heatmap or gaze plot on the observed medium, be it a computer screen or a physical environment.

Eye-tracking bridges the gap between the digital and real worlds by making virtual interactions more human-like. When social avatars in XR mimic real eye movements by blinking, gazing, and showing emotions, it deepens the sense of presence and immersion for the user.

Likewise, eye-tracking can drastically improve the understanding of user intent in the XR space. This means that XR environments can adapt in real time to the user's focus. For example, a horror game could trigger a scare only when the user is looking in the right direction, maximizing the emotional impact. By analyzing where users look, how often, and for how long, developers can glean valuable insights. This can guide design choices, ensure important elements capture attention, and refine user interfaces.

Eye-tracking paves the way for more intuitive user interfaces. For instance, instead of navigating through menus using clunky hand controllers, users can simply gaze at a menu option to select it.

Fortunately, you can enrich your XR scene with eye-tracking, regardless of whether your VR headset supports it or not. If this sparks your curiosity, follow along with the tutorial in the next section.

Setting up an XR scene to support eye- and head-tracking

The XR Interaction Toolkit supports eye- and head-tracking, enhancing user engagement in XR apps. While eye-tracking specifically captures where the user's eyes are focused, head-tracking determines the direction the user's head is pointing or gazing. Let's try out these tracking techniques ourselves by revisiting the basic VR scene we created in our last session.

Some key components of the XR Interaction Toolkit's eye- and head-tracking functionalities are already inside of our scene. To observe them, click on the arrow button next to the **XR Interaction Hands Setup** prefab in the **Scene Hierarchy** window of your project to see its children. Navigate to **XR Origin (XR Rig) | Camera Offset** and enable the **XR Gaze Interactor** and **Gaze Stabilized** prefabs. These prefabs are not only part of the **XR Interaction Hands Setup** prefab, but also the **XR Interaction Setup** prefab, which you would use in your scene if you don't include hand-tracking.

These prefabs work with all kinds of VR headsets, regardless of whether they support eye-tracking or not. If your headset doesn't support eye-tracking, it will use the built-in head-tracking feature of the VR headset to estimate the head gaze.

While incorporating eye- and head-tracking into our scene is a significant step forward, it's only part of the equation for crafting intuitive gaze-based interactions in XR. If it were that simple, there would be no need for this chapter, especially considering we've utilized this prefab in every VR scene throughout this book. To truly harness the capabilities of eye-tracking, we must also populate our scene with objects that can interact with the **XR Gaze Interactor** prefab. Let's create these objects:

1. To keep our scene organized, right-click on Hand Tracking Cube in the **Scene Hierarchy** window and select **Create Empty Parent**. Rename it Eye Tracking and Hand Tracking Interactables. This GameObject will store two more cubes enabling different kinds of eye-tracking interactions alongside the cube we already created.

2. Let's place all of the cubes on a very simple table. Create the table by clicking the + button in the **Scene Hierarchy** window and selecting **3D Object | Cube**. Rename the cube Table, position it at (0, 0.5, 0), and scale it to (3, 1, 1).

3. Scale `Hand Tracking Cube` to (`0.5, 0.5, 0.5`) and position it on `Table` by changing the values to (`-1, 0, 0`). Also, make sure you uncheck the **Is Kinematic** checkbox to let it obey the laws of gravity once more. Right-click `Hand Tracking Cube` and select the **Duplicate** option.

   Repeat the duplication process one more time so that you have created two new cubes alongside `Hand Tracking Cube`. Rename the two cubes `Eye Tracking Cube 1` and `Eye Tracking Cube 2`. Change the position of `Eye Tracking Cube 1` to (0, 0, 0) and the position of `Eye Tracking Cube 2` to (1, 0, 0).

4. Next, let's add the function that when the user's gaze is pointing at `Eye Tracking Cube 1`, a sphere will appear on top of it. We can add a sphere to our scene by selecting `Eye Tracking Cube 1` in the **Scene Hierarchy** window, right-clicking on it, and navigating to **3D Object | Sphere**. In the **Inspector** window of **Sphere**, position it at (0, 1, 0) and scale it to (`0.5, 0.5, 0.5`).

5. Similar to `Eye Tracking Cube 1`, some text should appear when the user of our VR scene looks at `Eye Tracking Cube 2`. To make the interaction more complex, the text should change, depending on whether the user looks at the cube and whether the cube is selected or deselected via a controller button press. To create the needed UI elements, right-click on `Eye Tracking Cube 2` in the **Scene Hierarchy** window and select **UI | Panel**. This will automatically add **Canvas** and **Panel** elements to your scene. Select **Canvas** and navigate to the **Inspector** window. Here, select the **World Space** option in the **Render Mode** drop-down menu of the **Canvas** component. Change its position to (0, `-1, -1.01`) and its **Width** and **Height** to 1. Now, right-click on **Panel** and select **UI | Text – TextMeshPro**. In the **Inspector** window of the newly created **Text** element, go to the **TextMeshPro – Text (UI)** component and insert `No state detected!` in the **Text Input** field. You should never see this text message if you follow this tutorial precisely, so regard it as a helpful addition to test your scene later. Feel free to change some of the other settings to adjust the look of the text. Rename the text element `Interactable Text`, scale it to `0.005` in all directions, and position it at (`0.1, 0.1, 0`).

6. Let's add some colorful materials to our scene to make it more visually appealing. Specifically, we want to add two materials for our cubes – one for their default appearance and one if they are hovered over. So, we will create a material for `Table` and one for **Sphere**. To accomplish this, right-click in the **Projects** window and select **Create | Folder**. Rename this folder `Materials` and double-click on it to open it. Create a new material inside the `Materials` folder by right-clicking, selecting **Create | Material**, and naming it `HighlightedCube`. In the **Inspector** window of `HighlightedCube`, click on the colored cell next to the **Base Map** property and select your preferred color for when the cube is highlighted. In our case, we only want the cube to become slightly brighter when it is highlighted, so we have chosen a yellow color (R: `240`, G: `240`, B: `140`) with an **Alpha** value of `70`.

7. Right-click inside the `Materials` folder and select **Create | Material** to create the `CubeMaterial`, `TableMaterial`, and `SphereMaterial` materials. For `CubeMaterial`, we chose a sophisticated red color (R: 186, G: 6, B: 6); `TableMaterial` has been assigned a dark brown color (R: 58, G: 40, B: 3); and for `SphereMaterial`, we selected a blue color (R: 29, G: 120, B: 241) and made it appear metallic by setting the **Metallic Map** slider to 1. All of these materials have an **Alpha** value of 255. Apply `CubeMaterial` to all three cubes in the scene by simply dragging and dropping the material from the **Project** window into the cubes in the scene. Do the same for `TableMaterial` and `SphereMaterial`.

Figure 8.3 shows what your VR scene should currently look like.

Figure 8.3 – The current status of the VR scene for eye-tracking

In the next section, you will learn how you can interact with these cubes via eye or head gaze.

Interacting with objects via eye and head gaze using the XR Interaction Toolkit

When a user looks at one of the two cubes we created for gaze-tracking, they should be able to interact with them as described in the previous section. To accomplish this, we need to adjust some components of our scene.

The **XR Gaze Interactor** prefab will only interact with objects in a scene that have either an **XR Simple Interactable** or **XR Grab Interactable** script attached to them and that have the **Allow Gaze Interaction** checkbox of the respective script enabled. This means we must add either one of these two scripts to our two new cubes.

In our case, we will add the **XR Simple Interactable** script to each of the two eye-tracking cubes. We need to perform the following three steps to achieve this:

1. Press *Ctrl/Cmd* and select both cubes in the **Scene Hierarchy** window.

2. Click the **Add component** button in the **Inspector** window, type XR Simple Interactable into the search bar, and double-click on the script to add it to both cubes.

3. Click the arrow next to the **Gaze Configuration** property of the **XR Simple Interactable** script component to open the gaze-related properties. Select the **Allow Gaze Interaction** checkbox.

Now, we need to specify how we can interact with our two cubes via gaze. Let's start with the first cube. Remember that our goal is to make the blue sphere appear on top of the first cube only when our eyes, head, or controllers are directed toward it. We can add this logic to the cube by following these steps:

1. In the **Inspector** window of Eye Tracking Cube 1, navigate to the end of the **XR Simple Interactable** script component. Click on the arrow next to **Interactable Events** to open it. Add two new events to the **First Hover Entered** function of **First/Last Hover** by clicking on the + button two times.

2. Let's fill in the missing information of our newly created interactable events. As indicated by the **None (Object)** cell, neither event is currently assigned to a GameObject. To change this, drag and drop Eye Tracking Cube 1 from the **Scene Hierarchy** window into the first **None (Object)** cell. Repeat this for the second event by assigning it to **Sphere**.

3. To change the material of Eye Tracking Cube 1 when it is hovered over via controllers, the eyes, or the head, we must assign the necessary functions to each interactable event and provide them with the needed parameters. Select **MeshRenderer | Material material** for the **No Function** cell of the first event. A new cell will appear where we can assign the material the cube should have once it is hovered over. Click on the rounded dot icon next to the new cell. By doing so, a new window will open, where you can search for the HighlightedCube material via the search bar and select it. For the **No Function** cell of the second event, select **GameObject | SetActive (bool)** and be sure that the newly appeared checkbox is checked. This ensures that when you hover over Eye Tracking Cube 1 with your controllers, eyes, or head, you will see the blue sphere appearing on top of it.

4. Once your controllers, eyes, or head are no longer directed toward Eye Tracking Cube 1, the cube's material should change back to its default color and the sphere should disappear. To accomplish this, go to the **Last Hover Exited** function of **First/Last Hover** and press the + icon two times to create two new interactable events. Once again, drag and drop Eye Tracking Cube 1 and **Sphere** from the **Scene Hierarchy** window into the **None (Object)** cells to assign both events to their respective GameObjects. Assign the **MeshRenderer | Material material** function to the first event by clicking on the **No Function** cell. To change the cube's color back to red, press the dot icon next to the newly appeared cell, search for CubeMaterial, and select it. Repeat this process for **Sphere** by assigning it to the **GameObject | SetActive (bool)** function again and ensuring the newly appeared checkbox is not checked this time.

5. To evaluate the logic we've put in place so far, it's necessary to either disable or hide the blue **Sphere** object in our scene. This ensures that upon initial entry into the scene, the user doesn't see it. You can achieve this by selecting **Sphere** within the **Scene Hierarchy** window and then deselecting the checkbox next to its name at the top of the **Inspector** window.

Voilà – we've completed all the required steps to add a powerful eye-tracking capability to our first cube. *Figure 8.4* shows what the cube should look like once your eyes or head are directed toward it.

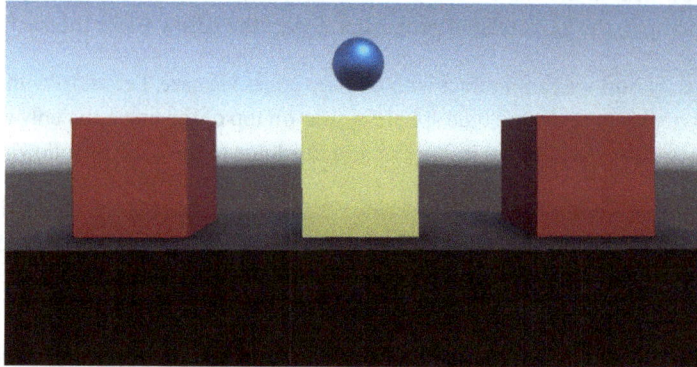

Figure 8.4 – The highlighted cube and blue sphere appear because
of your eyes or head hovering over the cube

If you are working with head gaze, you will only see the blue sphere when the cube is positioned at the center of your current field of view when you are wearing a VR headset or using the XR Device Simulator. You can observe the same effect when your controllers are pointing at the cube.

We will add an even more powerful interaction to our second cube. Like the first cube, a text element will appear underneath the second cube when the eyes, head, or controllers are directed toward it. This time, however, the displayed text itself will change based on whether the cube is hovered over, selected, or deselected using a combination of eye, head, and controller interactions. Follow these steps to accomplish this goal:

1. Repeat all the steps you performed for `Eye Tracking Cube 1` with the following modification: drag and drop `Eye Tracking Cube 2` and **Canvas** into the two events of the **First Hover Entered** function and the **Last Hover Exited** function.

2. So far, the cube's color changes and the text element becomes visible once the cube is hovered over. However, the text itself should also change to **Hovered** once this event is triggered. To accomplish this, add a third event to the **First Hover Entered** function by clicking the + icon. Drag and drop `Interactable Text` from the **Scene Hierarchy** window into the third event's **None (Object)** cell. Assign the **TextMeshProUGUI | string text** function to this event and type `Hovered` into the newly appeared empty text cell.

3. When the cube is hovered over and selected by pressing the controller button, which is typically reserved for moving objects around, the text displayed underneath the cube should change. To implement this logic, scroll down to **First/Last Select** and click on the + icon of the **First Select Entered** function. Drag and drop the `Interactable Text` element from the **Scene Hierarchy** window into the **None (Object)** cell and assign the **TextMeshProUGUI | string text** function to it. Type the word `Selected` into the empty text field.

4. Once the cube is no longer selected via controllers, the text should change again. This can be accomplished by adding a new event to the **Last Select Exited** function of **First/Last Select**. After clicking the + icon, assign `Interactable Text` to the **None (Object)** cell per drag-and-drop and, as before, choose **TextMeshProUGUI | string text** and insert `Deselected` into the newly created text cell.

5. Before testing out the logic of the second cube, hide it in the scene by deselecting it in its **Inspector** window, as you did with the first cube.

It is time to try out interacting with this cube. *Figure 8.5* shows the cube in its three interactable states – being hovered over, being hovered over plus selected, and being hovered over but deselected.

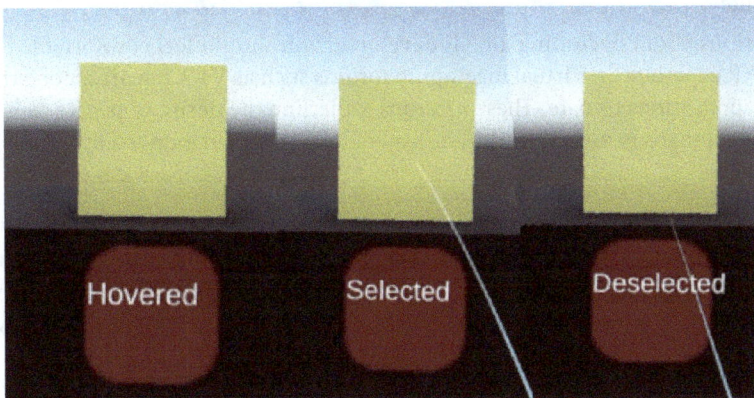

Figure 8.5 – The cube in its three interactable states

Hurray – you have successfully added different eye-tracking interactions to your scene! In the next section, you will learn how to set up a multiplayer XR game in Unity.

Building a VR multiplayer application

In this section, you'll craft your first VR multiplayer application. Though it involves a fair amount of C# coding, you've gained enough knowledge to smoothly navigate this tutorial. But before we dive in, let's pause and delve into the nuances of constructing multiplayer applications, the necessary components, and the appeal of multiplayer experiences within the XR context.

Understanding multiplayer applications and multiplayer networking systems

At its core, a **multiplayer** experience allows multiple users to interact within a shared digital environment simultaneously. This environment can range from simple text-based interfaces to complex virtual realities. The key components of a multiplayer experience typically include servers that host the game environment, networking systems that handle data synchronization and communication, player avatars, and game logic that governs interaction rules.

Adding a multiplayer mode to a game can make it more unpredictable due to human behaviors and decision-making. By contrast, **single-player** modes are typically more controlled and can be designed around a predefined narrative.

By adding multiplayer capabilities to XR, users are encouraged to jointly engage in tasks, challenges, or experiences, making the environment feel more alive and dynamic. Examples include cooperative puzzle-solving, virtual team-building exercises, and joint exploration.

> **Important note**
>
> If you are serious about becoming an XR developer, you should feel comfortable creating multiplayer XR experiences. Virtual meetups in forums such as VR Chat are among the most popular XR applications to date. These meetups underline the desire of people to hang out with other real humans in virtual worlds, instead of being only surrounded by virtual assets.

The most crucial part of any multiplayer game is the **multiplayer networking system**. At its essence, a multiplayer networking system is the digital backbone that enables various players to interact seamlessly within a shared environment. This system ensures that actions performed by one player are reflected accurately and consistently for all other players, creating a harmonized virtual experience.

A key component of every multiplayer networking system is **servers**. These are powerful computers that host the game's digital environment. They are the central point that players connect to, and they maintain the authoritative state of the game. In some configurations, one of the players might act as a server, termed **peer-to-peer**, but dedicated servers are more common in larger games due to stability and scalability.

The individual devices or computers used by players are called **client systems**. They send data such as player movements or actions to the server and receive updates about the game world and other players.

In real-time games, even slight delays can affect gameplay. Multiplayer networking systems use techniques such as **lag compensation** to make sure players have smooth experiences. This involves predicting movements or actions and then reconciling differences once actual data arrives.

To avoid multiplayer games becoming targets for cheating or hacking, networking systems employ measures such as data encryption, authoritative servers, and cheat detection tools.

Beyond gameplay, players often wish to communicate, be it through text, voice, or other mediums. Networking systems provide the infrastructure for these interactions, ensuring real-time and clear communication.

Here is an overview of the three main multiplayer network providers that are interesting for Unity developers:

- **Photon Unity Networking (PUN)**: PUN is a solution tailored for Unity's multiplayer games. Its cloud-based approach means that developers don't need to worry about server creation or maintenance. PUN offers a free tier. Although it comes with certain restrictions, the free version is heavily used by indie developers and hobbyists who are starting their journey into multiplayer game development.

- **Mirror**: Mirror offers a community-powered, open source networking tool tailored for Unity. It's an evolution of the deprecated Unity networking system. Being open source, it doesn't impose licensing fees, making it an economical choice. Mirror is renowned for its flexibility and provides developers with a greater degree of control over their multiplayer logic. However, its customizable nature means that it presents a slightly more challenging learning curve for beginners.

- **Netcode**: Netcode is Unity's proprietary solution for game networking, formerly known as *MLAPI*. This evolution and rebranding signifies Unity's commitment to continuous improvement and its dedication to providing developers with top-tier tools for multiplayer game development. Being an intrinsic Unity solution, Netcode promises seamless integration with Unity's ecosystem.

In the following sections, we'll create a multiplayer game of our own. For this, we'll be leveraging Photon PUN 2 using the free tier. Its zero-cost barrier, combined with an intuitive setup process for beginners, makes it an ideal candidate for rapid prototyping and smaller-scale projects.

The next section will teach you how to set up PUN for the VR multiplayer game we are about to create.

Setting up PUN for our VR multiplayer game

Our objective for this VR multiplayer game is both straightforward and robust, encompassing all essential elements needed for a multiplayer experience. We aim to design a VR scene where multiple users can join concurrently. Users should see themselves and each other through avatars that consist of a head and two controller components. They should witness the movements of others in this shared space in real time, observing how they maneuver or rotate their controllers. Moreover, through hand animations, they should be able to discern when others are pressing their hands.

Before we start importing Photon PUN 2 into our project, let's create a new scene for this part of this chapter. Follow these steps to get started:

1. Open the `AdvancedXRTechniques` project, which we created in the previous sections of this chapter. Save the current scene using an expressive name such as `HandAndEyeTrackingScene` by navigating to **File** | **Save As** and typing in the scene name of your choice.

2. Go to **File | New Scene** and select the **Standard (URP)** template to create a new scene. Save this scene with a meaningful name too by going to **File | Save As** and inserting the name MultiplayerScene.

3. In the Unity Editor of your new scene, delete **Main Camera** from the **Scene Hierarchy** window by right-clicking on it and selecting **Delete**. Instead of **Main Camera**, we need to add the **XR Interaction Setup** prefab to our scene. To do this, search for XR Interaction Setup via the search bar of the **Project** window, select it, and drag the prefab into the **Scene Hierarchy** window. Make sure **XR Interaction Setup** is positioned at (0,0,0).

Now that we've set up our scene, follow these steps to install and set up our networking system, PUN:

1. We can install PUN like any other package via the Unity Asset Store. Visit https://assetstore.unity.com/packages/tools/network/pun-2-free-119922 or search for PUN 2 - FREE via the Unity Asset Store (**Window | Asset Store**). Click the **Add to my Assets** button to add the package to your assets. Once the package has been added, a new button called **Open in Unity** will appear on the Asset Store's website. Click it to open the package in the **Package Manager** window of your project. Press the **Download** and **Import** buttons to import the asset into your project.

2. The **PUN Wizard** window shown in *Figure 8.6* will pop up in the Unity Editor.

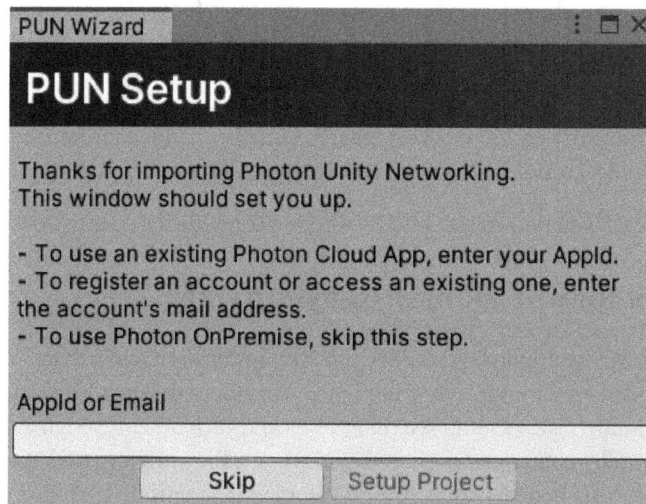

Figure 8.6 – The PUN Wizard pop-up window

If you already have a Photon account, you can simply type in your **AppId** here. Otherwise, input your email and click on the **Setup Project** button once you are done. This will create an account and forward you to a web page where you can set your password.

> **Important note**
> If you accidentally close this pop-up window, simply head to https://www.photonengine.com/ and sign up there.

3. Once you have signed in to your account, head to https://dashboard.photonengine.com/ and click on the **Create A New App** button. You will be forwarded to the page shown in *Figure 8.7*.

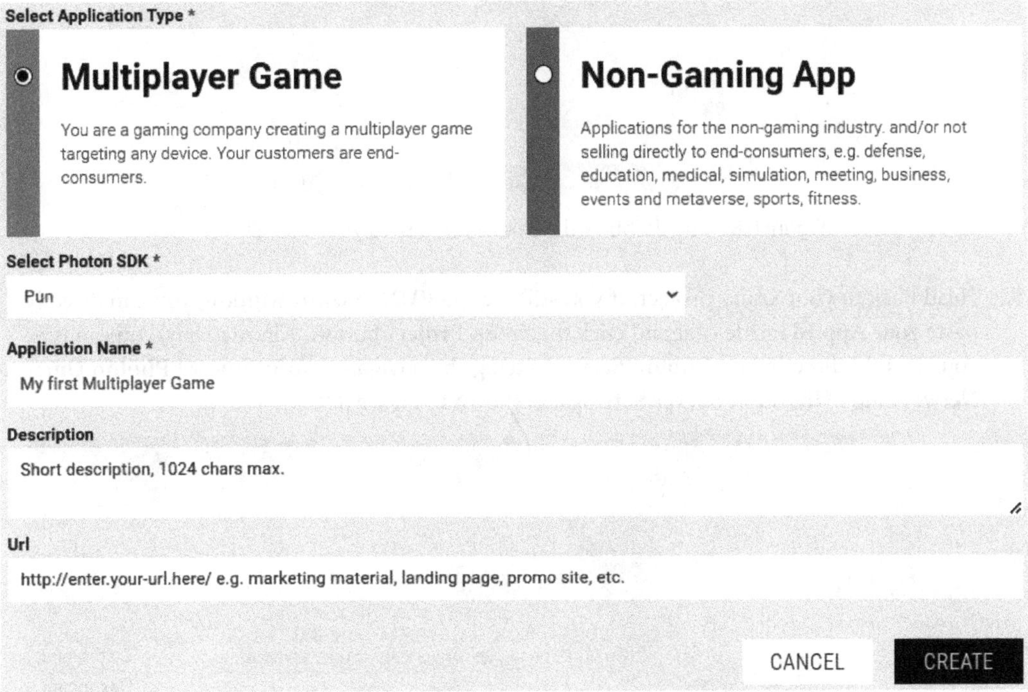

Select Application Type *

⊙ **Multiplayer Game**

You are a gaming company creating a multiplayer game targeting any device. Your customers are end-consumers.

○ **Non-Gaming App**

Applications for the non-gaming industry. and/or not selling directly to end-consumers, e.g. defense, education, medical, simulation, meeting, business, events and metaverse, sports, fitness.

Select Photon SDK *

Pun ⌄

Application Name *

My first Multiplayer Game

Description

Short description, 1024 chars max.

Url

http://enter.your-url.here/ e.g. marketing material, landing page, promo site, etc.

CANCEL   CREATE

Figure 8.7 – The web page you will be forwarded to once you
create a new application on the Photon website

4. Choose the **Multiplayer Game** option under **Select Application Type**, select **Pun** from the **Select Photon SDK** drop-down menu, and insert an **Application Name** value of your choice, such as My first Multiplayer Game. Click the **CREATE** button – notice your newly created application in the dashboard. As shown in *Figure 8.8*, you can find your Photon **App ID** in one of the layout elements of the dashboard. Copy it to connect to the server later.

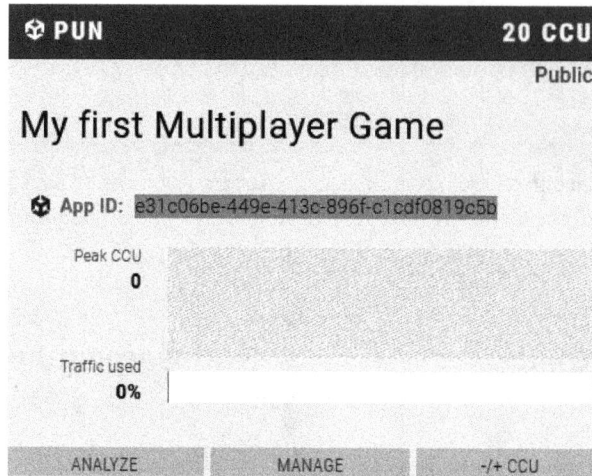

Figure 8.8 – The dashboard element containing your App ID

5. Head back to your Unity project. If you still see the **PUN Wizard** window, you can directly paste your **App Id** inside of it and click the **Setup Project** button. Alternatively, you can type your **App Id** directly into **Photon Server Settings** by navigating to **Window | Photon Unity Networking | Highlight Server Settings**, as shown in *Figure 8.9*.

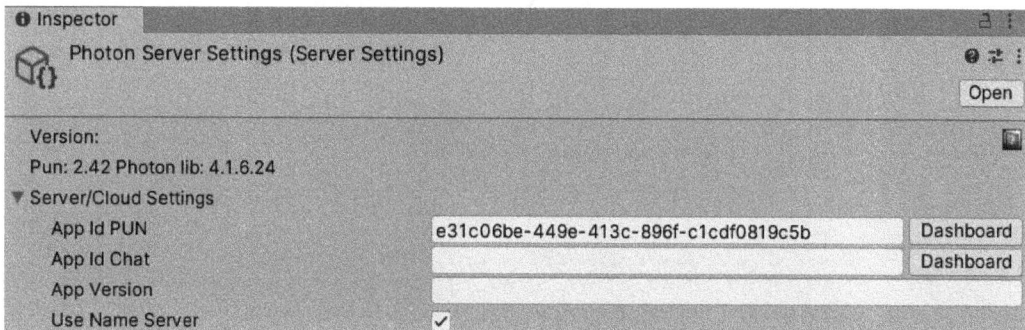

Figure 8.9 – How to add the App Id to Photon Server Settings in Unity

Now, it's time to connect our Unity project to the server. The next section will show you how to do this.

Connecting to the server via Network Manager

To connect our Unity project to the PUN server, we must add a new GameObject with an associated C# script to our scene. Let's go through this process step by step:

1. Create a new empty GameObject by right-clicking in the **Scene Hierarchy** window, selecting **Create Empty**, and renaming it to Network Manager.

2. Now, we want to add a script to this GameObject that we can use to connect to the PUN server and check whether someone else joined the server. To create this script, navigate to the **Inspector** window of Network Manager and click the **Add Component** button.

3. Search for NetworkManager, select the **New Script** option, and press the **Create and Add** button to create a C# script called **NetworkManager.cs** that is automatically added to **NetworkManager** as a component. Double-click on the script to open it in your preferred IDE.

4. The **NetworkManager** script serves as a foundational element for initiating and managing network interactions in a VR multiplayer application using the PUN framework. Delete everything that is currently inside of the **NetworkManager** script so that you can start on a clean foundation.

Let's start adding our code logic by importing the following libraries:

```
using UnityEngine;
using Photon.Pun;
using Photon.Realtime;
```

The Photon.Pun and Photon.Realtime libraries are vital for any PUN application as they grant access to core multiplayer networking functionalities.

Next, let's define the NetworkManager class and some important constants:

```
public class NetworkManager : MonoBehaviourPunCallbacks
{
    private const string ROOM_NAME = "Multiplayer Room";
    private const byte MAX_PLAYERS = 5;
}
```

The NetworkManager class inherits from MonoBehaviourPunCallbacks. This inheritance means that it's not just a standard Unity script (MonoBehaviour), but that it also has special callback functions provided by PUN that notify our script of various networking events.

While ROOM_NAME represents the name of the room players will join or create, MAX_PLAYERS defines the maximum number of players allowed in a room, which in this case is 5 players. We will need both constants later in the script.

Let's continue by connecting our application to Photon servers:

```
private void Awake()
{
    InitiateServerConnection();
}
private void InitiateServerConnection()
{
    if (!PhotonNetwork.IsConnected)
    {
```

```
        PhotonNetwork.ConnectUsingSettings();
        Debug.Log("Attempting server connection...");
    }
}
```

Once the `Awake()` method is invoked, it initiates the connection to the PUN server by calling the `InitiateServerConnection()` method. If the client isn't already connected to Photon, the `InitiateServerConnection()` method will attempt to connect to the server space using the **App Id** value that we inserted before. Refer to *Figure 8.9* in case you missed this step. The second line of this method logs our attempt via a debug message.

If the connection attempt is successful, the following method will be executed:

```
public override void OnConnectedToMaster()
{
    base.OnConnectedToMaster();
    Debug.Log("Connected to Master Server.");
    JoinOrCreateGameRoom();
}
```

This method is an override of the `OnConnectedToMaster` callback from Photon. It is triggered once the application successfully connects to the Photon Master Server. This method is an integral part of the PUN framework, allowing us to execute specific logic after a successful connection. Inside this method, we log the successful connection to the Master Server. We also call the `JoinOrCreateGameRoom()` method, which uses the constants we defined at the beginning of our script:

```
private void JoinOrCreateGameRoom()
    {
        RoomOptions options = new RoomOptions
        {
            MaxPlayers = MAX_PLAYERS,
            IsVisible = true,
            IsOpen = true
        };

        PhotonNetwork.JoinOrCreateRoom(ROOM_NAME, options, TypedLobby.
Default);
    }
```

In the last line of this method, the client attempts to join or create a room, called **Multiplayer Room**, with five being the maximum allowed number of players in the room. If the room doesn't exist, it creates one with the provided options.

The next method in our code is an override:

```
public override void OnJoinedRoom()
{
    base.OnJoinedRoom();
    Debug.Log("Successfully joined a room.");
}
```

This method is an override of the `OnJoinedRoom` callback from Photon. It's triggered when the client successfully joins a room. In the last line, a debug message is printed to confirm successful room entry.

To manage new players, we can use the following method:

```
public override void OnPlayerEnteredRoom(Player newParticipant)
{
    base.OnPlayerEnteredRoom(newParticipant);
    Debug.Log("Another player has joined the room.");
}
```

This `OnPlayerEnteredRoom()` method is also an override of the `OnPlayerEnteredRoom` callback from Photon. It's called whenever a new player enters the room. Once again, a debug message is printed in the last line to notify that another player has joined.

Hurray, you have implemented all the necessary components into the `NetworkManager` script! As you can see, it provides a foundational structure to connect to Photon's servers, manage multiplayer rooms, and handle player interactions in a VR environment. Let's see whether our code logic works. In the next section, you will find out how you can test a multiplayer application on a single device.

Testing the multiplayer scene from one device

To test a multiplayer scene from a single device, we need to run the scene twice. This can be accomplished by first building and running the scene on the computer and subsequently launching it from within the Editor. Here's a step-by-step breakdown:

1. In the Unity Editor, head to **Files | Build Settings** and select the **Windows/ Mac/ Linux** tab. Add the open scene you want to test, click the **Build** button, and choose a folder for the executable file you are about to build.

2. Once the file has been built, click on the **Play** button in the Unity Editor to start the scene.

3. Now, open the executable file that you just built.

4. Head to the **Console** window in the Unity Editor and check the **Debug** statements. If you did everything correctly, you will see the debug lines we defined in the scripts previously, as shown in *Figure 8.10*.

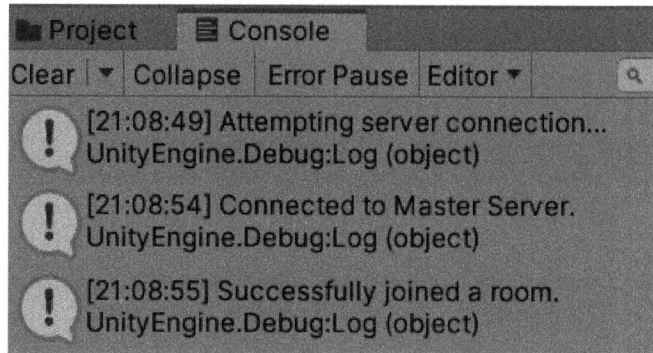

Figure 8.10 – The debug lines you should see if everything works as expected

To ensure that our scene functions optimally, we must test its VR features thoroughly. From our assessments, navigating the scene is smooth, and both the headset and controllers are accurately tracked and positioned. Currently, our scene establishes a server connection as soon as the initial player activates it, and we have set up methods to detect the entry of new players.

However, our **XR Interaction Setup** isn't entirely primed for multiplayer operations. It excels at managing locomotion and interactions, but it doesn't feature animated models for different body parts. In simpler terms, if someone were to enter our multiplayer scene at this moment, they might only see the controller. Worse, they might not detect any part of our avatar because the controller models aren't network-conscious. This means that the objects present aren't synchronized across all players in the session. But don't worry, we'll tackle this issue in the upcoming section.

Using scripts to display our avatar's hands and face

Our next objective is to display an avatar's hands and face when it connects to a server. To achieve this, let's append a new script to Network Manager by selecting it in the **Inspector** window, searching for NetworkPlayerPlacer after clicking the **Add Component** button in the **Inspector** window, and creating a new script called NetworkPlayerPlacer. Open the script by double-clicking on it.

Managing player presence

The NetworkPlayerPlacer script we are about to dive into is tailored for overseeing player presence, a core element in multiplayer VR applications. Specifically, this means to create and delete player avatars or their in-game representations. The NetworkPlayerPlacer script is tailored for this exact role. It begins with the following declarations:

```
using UnityEngine;
using Photon.Pun;
public class NetworkPlayerPlacer : MonoBehaviourPunCallbacks
{
```

```
    private GameObject playerInstance;
    private const string PLAYER_PREFAB_NAME = "Network Player";
}
```

The import statements and class declarations bear a close resemblance to those in the `NetworkManager` script. Both classes are derived from `MonoBehaviourPunCallbacks`. `playerInstance` serves as a reference to the instantiated player object within the scene. Meanwhile, `PLAYER_PREFAB_NAME` is a constant string that holds the name of the player prefab set for instantiation. It's anticipated that this prefab is registered and accessible in the Photon resources directory.

Let's create the first method of the `NetworkPlayerPlacer` script:

```
public override void OnJoinedRoom()
{
    base.OnJoinedRoom();
    SpawnPlayer();
}
```

The `OnJoinedRoom()` method overrides the `OnJoinedRoom` callback from Photon, which gets triggered when the local player successfully joins a room.

The base class implementation of `OnJoinedRoom` is called with `base.OnJoinedRoom()`. In the last line, the `SpawnPlayer()` method is called, which looks like this:

```
private void SpawnPlayer()
{
    playerInstance = PhotonNetwork.Instantiate(PLAYER_PREFAB_NAME,
transform.position, transform.rotation);
}
```

This method instantiates a new player object using Photon's networked instantiation method. The new player will be spawned at the position and rotation of the `NetworkPlayerPlacer` object. This ensures that the player object is networked and synchronized across all clients in the room.

The next method in our script overrides the `OnLeftRoom` callback from Photon, which is called when the local player leaves a room:

```
public override void OnLeftRoom()
{
    base.OnLeftRoom();
    DespawnPlayer();
}
```

The `OnLeftRoom()` method consists of two calls: one to the base class implementation of `OnLeftRoom` and one to `DespawnPlayer()`, a method to despawn the player. Let's have a look at the `DespawnPlayer()` method next:

```
private void DespawnPlayer()
{
    if (playerInstance)
    {
        PhotonNetwork.Destroy(playerInstance);
    }
}
```

This method handles the despawning of the player object.

The `if` statement checks whether the `playerInstance` reference is not null. If this is true, a player object exists. The `PhotonNetwork.Destroy(playerInstance);` line destroys the networked player object. This will ensure that the object is removed not just locally but across all clients.

In essence, the `NetworkPlayerSpawner` script, with its two key callback functions, `OnJoinedRoom()` and `OnLeftRoom()`, seamlessly handles the appearance and disappearance of player avatars in our VR multiplayer space, complementing the functionalities offered by the `NetworkManager` script.

Creating a face and hands

Earlier, we noted that the script anticipates a prefab called **Network Player** in the **Project** window's `Resources` folder. Unity uses this default naming convention when referencing objects by their name. To create the needed folder, click on the + icon in the **Project** window, select **Folder**, and then rename it to `Resources`.

Next, we need to create the **Network Player** prefab, which will represent our avatar via the network. Follow these steps to achieve this:

1. Right-click in the **Scene Hierarchy** window, select **Create Empty**, and rename the empty GameObject to `Network Player`.

2. Add the **Photon View** component to it in the **Inspector** window by pressing the **Add Component** button and searching for it. Photon requires this component to instantiate the player on the server. Without it, the system wouldn't recognize ownership or synchronize the player's body parts accurately.

3. Add three other empty GameObjects as children to `Network Player` by selecting `Network Player` in the **Scene Hierarchy** window and pressing **Create Empty**. Repeat this process two times so that you have created three empty GameObjects in total and rename them to `Head`, `Left Hand`, and `Right Hand`.

4. Now, we need to create 3D objects to represent the three components of our avatar. For Head, we can use a simple **Sphere** primitive. Right-click on Head in the **Scene Hierarchy** window and select **3D Object | Sphere**. Rename it Head and scale it to (0.1, 0.1, 0.1)

5. We could also use primitives for the hands of our avatar. However, since there are a lot of animated hand models available on the internet, we can simply use one of them. For this project, we used the **Oculus Hands** sample from the **Oculus Integration** package on Unity's Asset Store (https://assetstore.unity.com/packages/tools/integration/oculus-integration-82022). However, we suggest cloning the hands directly from this book's GitHub repository. This avoids unnecessary additional content and missing assets due to frequent updates in the **Oculus Integration** package.

6. Now, select Network Player in the **Scene Hierarchy** window and drag it into the Resources folder in the **Project** window. This will create a prefab of it that we will instantiate at runtime. Hence, we don't need Network Player in our **Scene Hierarchy** window anymore and can delete it.

We must add another script to Network Manager to make sure the Head, Left Hand, and Right Hand components follow the position of the user's headset and controllers. To do this, select Network Manager in the **Scene Hierarchy** window, click the **Add component** button in the **Inspector** window, type NetworkPlayer into the search bar, and create a new script with this name.

Now that our script has been created, let's learn how we can use it to track the player's position and movement.

Tracking player position and movements

The NetworkPlayer script controls how a player's movements are tracked and represented within the multiplayer VR environment. The script encapsulates functionalities that ensure the position and rotation of the avatar's head, left hand, and right hand are accurately tracked and updated.

Let's have a look at the beginning of this script:

```
using UnityEngine;
using UnityEngine.XR;
using Photon.Pun;
using UnityEngine.InputSystem;
public class NetworkPlayer : MonoBehaviour
{
    public Transform head;
    public Transform leftHand;
    public Transform rightHand;
    private PhotonView photonView;

    public InputActionAsset xriInputActions;
    private InputActionMap headActionMap;
```

```
        private InputActionMap leftHandActionMap;
        private InputActionMap rightHandActionMap;

        private InputAction headPositionAction;
        private InputAction headRotationAction;
        private InputAction leftHandPositionAction;
        private InputAction leftHandRotationAction;
        private InputAction rightHandPositionAction;
        private InputAction rightHandRotationAction;
}
```

Besides importing the regular Unity and PUN namespaces, a class named `NetworkPlayer` inheriting from `MonoBehaviour` is declared. The `Transform` variables, `head`, `leftHand`, and `rightHand`, represent the 3D position, rotation, and scale of a player's VR avatar components in the game. The `PhotonView` component is fundamental for PUN, determining ownership and synchronization of objects in multiplayer. `InputActionAsset`, called `xriInputActions`, is a Unity-configured set of input definitions tailored for VR headsets. The `InputActionMap` variables, such as `headActionMap`, group related input actions, allowing for organized handling of inputs such as head or hand movements. Finally, the `InputAction` variables such as `headPositionAction` and `headRotationAction` detect individual input movements from the VR headset. To synchronize the player's real-world VR movements with their in-game avatar, we need methods in our script. Let's start with the first one:

```
void Start()
    {
        photonView = GetComponent<PhotonView>();

        // Get the Action Maps
        headActionMap = xriInputActions.FindActionMap("XRI Head");
        leftHandActionMap = xriInputActions.FindActionMap("XRI
LeftHand");
        rightHandActionMap = xriInputActions.FindActionMap("XRI
RightHand");

        // Get the Position and Rotation actions for each action map
        headPositionAction = headActionMap.FindAction("Position");
        headRotationAction = headActionMap.FindAction("Rotation");

        leftHandPositionAction = leftHandActionMap.
FindAction("Position");
        leftHandRotationAction = leftHandActionMap.
FindAction("Rotation");

        rightHandPositionAction = rightHandActionMap.
FindAction("Position");
```

```
        rightHandRotationAction = rightHandActionMap.
FindAction("Rotation");

        // Enable actions
        headPositionAction.Enable();
        headRotationAction.Enable();
        leftHandPositionAction.Enable();
        leftHandRotationAction.Enable();
        rightHandPositionAction.Enable();
        rightHandRotationAction.Enable();
    }
```

In the `Start()` method, `photonView` is initialized. Then, action maps of the head, left hand, and right hand are retrieved using their names. These names (`XRI Head`, `XRI LeftHand`, and `XRI RightHand`) are predefined in **XR Interaction Toolkit Starter Assets**. You will learn more about them later in this chapter. From these action maps, the individual `Position` and `Rotation` actions for the head and each hand are retrieved. Lastly, all these actions are enabled so that they begin reading input values.

The next method in our script is the `Update()` method:

```
void Update()
    {
        if (photonView.IsMine)
        {
            rightHand.gameObject.SetActive(false);
            leftHand.gameObject.SetActive(false);
            head.gameObject.SetActive(false);

            MapPosition(head, XRNode.Head);
            MapPosition(leftHand, XRNode.LeftHand);
            MapPosition(rightHand, XRNode.RightHand);
        }
    }
```

In a Photon-powered multiplayer environment, numerous `NetworkPlayer` instances will emerge, representing each player in the game. However, on every player's device, only a single instance genuinely represents that specific player, while the rest signify remote players. The `photonView.IsMine` property, when checked within the `Update()` method's `if` statement, determines whether the `NetworkPlayer` instance on a given machine corresponds to the player using that machine. This differentiation is pivotal in multiplayer scenarios to identify who has control over certain activities or decisions. If it returns true, then the `NetworkPlayer` instance pertains to the local player of that device. Consequently, the visual representation of that player's head and hands is deactivated, ensuring a seamless user experience.

Within our multiplayer scene, two primary players exist: our local player (via **XR Interaction Setup**) and the remote player (`Network Player`). This remote player is visible to all other participants in the multiplayer environment. As the local player already has hand models, introducing an additional hand model from the remote player can lead to a disorienting experience, particularly given network update latencies, as you can see in *Figure 8.11*.

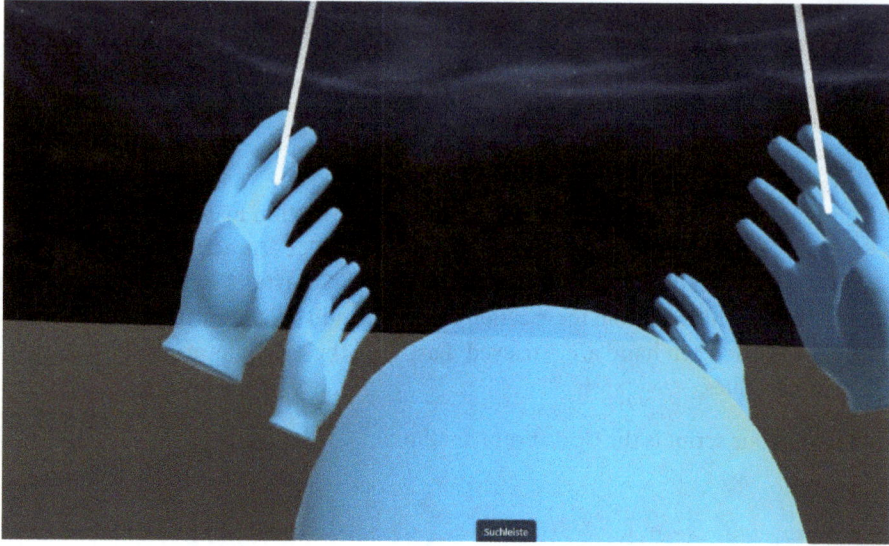

Figure 8.11 – What happens when the SetActive(false) statements are not implemented

Thus, we ensure the elements of `Network Player` remain invisible to the user. Conversely, if the property returns `false`, it means the instance represents another participant as viewed from that device. Finally, the `if` statement also updates the position and rotation of the hands and head based on VR headset or VR controller movements using the `MapPosition()` method.

Now, let's have a look at the `MapPosition()` method:

```
void MapPosition(Transform target, XRNode node)
    {
        Vector3 position = Vector3.zero;
        Quaternion rotation = Quaternion.identity;

        if (node == XRNode.Head)
        {
            position = headPositionAction.ReadValue<Vector3>();
            rotation = headRotationAction.ReadValue<Quaternion>();
        }
        else if (node == XRNode.LeftHand)
        {
```

```
                position = leftHandPositionAction.ReadValue<Vector3>();
                rotation = leftHandRotationAction.ReadValue<Quaternion>();
        }
        else if (node == XRNode.RightHand)
        {
                position = rightHandPositionAction.ReadValue<Vector3>();
                rotation = rightHandRotationAction.
ReadValue<Quaternion>();
        }

        target.position = position;
        target.rotation = rotation;
    }
```

The purpose of the `MapPosition()` method is to assign the correct position and rotation to the passed-in **Transform** based on the XR input. It starts by initializing the position and rotation to default values. Then, it checks whether the passed-in node is the head, left hand, or right hand via an `if` statement and reads the corresponding input values. For example, if the node is `XRNode.Head`, it reads the position and rotation values from `headPositionAction` and `headRotationAction`, respectively. These `InputAction` variables fetch the latest position and rotation values from the VR headset at every frame. Finally, the `MapPosition()` method assigns the fetched position and rotation to the passed-in target transform.

In essence, this method ensures that the in-game representation of the player moves in sync with their real-world movements in VR.

The last method in this script is the `OnDestroy()` method:

```
void OnDestroy()
    {
            headPositionAction.Disable();
            headRotationAction.Disable();
            leftHandPositionAction.Disable();
            leftHandRotationAction.Disable();
            rightHandPositionAction.Disable();
            rightHandRotationAction.Disable();
    }
```

The `OnDestroy()` method simply disables all the input actions when the script is destroyed, ensuring no unwanted input readings are occurring in the background.

Overall, the `NetworkPlayer` script serves as a bridge, translating a player's real-world VR movements into the virtual multiplayer space. By adding this script to our scene, players can experience a synchronous and immersive VR multiplayer environment.

When we test the application now, we can see the head and both of the hands somewhere in the scene aligned with our controller position. Next, we'll animate our hands when we press the trigger and grip button.

Animating the hand models

Your multiplayer application has come a long way. Now, let's take the immersion one step further with hand animations. By mapping controller inputs to these animations, users will experience a heightened realism within the virtual world. If animations sound foreign to you, we recommend a quick detour to the *Understanding animations and animator systems* section of *Chapter 5*.

Within the **Project** window, navigate to **Multiplayer Scene | Oculus Hands**. To breathe life into our hands, we need to head to the Prefabs folder and then follow these steps:

1. Attach an **Animator** component to the hand prefabs by clicking the **Add component** button in the **Inspector** window and searching for Animator.

    In *Chapter 5*, we touched on **Animator** in Unity, which is essential for controlling character animations. **Animator** needs **Animator Controller** to manage how animations transition and interact. It also requires **Avatar**, which is a map of the character's skeletal structure, ensuring animations fit and move realistically on the character.

2. For our avatars, we can select the **L_hand_skeletal_lowres** and **R_hand_skeletal_lowres** options from **Multiplayer Scene | Oculus Hands | Models**.

3. However, we're yet to acquire **Animator Controllers**. We need one for each hand. Initiate this by heading to **Multiplayer Scene | Oculus Hands | Animations** in the **Project** window. Subsequently, right-click and select **Create | Animator Controller**. Repeat this and name them Left Hand Animator and Right Hand Animator.

Double-click on Left Hand Animator to reveal an **Animator** window showcasing **Layers** and a **Parameters** tab. Start with the **Parameters** tab and introduce **Grip** and **Trigger** parameters, both of the *float* type. They indicate the pressure intensity of the trigger and grip actions:

*   **Trigger** determines the degree of the hand's pinch
*   **Grip** governs the degree of the fist's bend

However, simply having these parameters won't magically animate the fingers. To bring them to life, we need to bind these parameters to specific animation states or integrate them within blend trees. These concepts have different approaches to connecting animation states, so let's compare them to see which approach fits our needs the best:

*   **Transitions** form the connection between varying animation states. For illustration, if you've defined two animation states, say **HandOpen** and **HandClosed**, the **Grip** parameter can dictate the switch between these. For example, if **Grip** surpasses 0.5, the hand might transition to **HandClosed**; otherwise, it will remain in the **HandOpen** state.

- **Blend trees** enable seamless transitioning between multiple animations depending on one or more parameters. It's analogous to using a music mixer to blend songs.

For hand animations, it is useful to have three states: **HandOpen**, **HandHalfClosed**, and **HandFullyClosed**. Instead of basic transitions, which can feel abrupt, a blend tree can be used to fluidly move between these states, all based on the **Grip** value. This is why it is the best choice for our needs.

To set up a blend tree, right-click in the **Base Layer** window's center and opt for **Create State | From New Blend Tree**. Your **Baser Layer** window should be similar to what's shown in *Figure 8.12*.

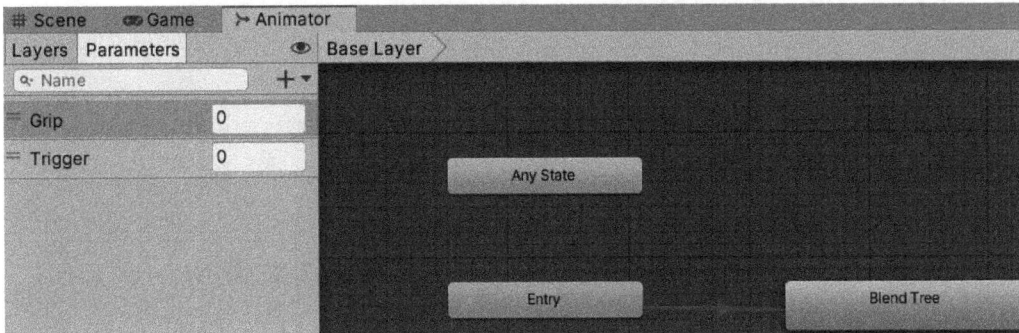

Figure 8.12 – Base Layer in the Animator window

By double-clicking on **Blend Tree** and selecting it afterward, you'll be prompted to select a **Blend Type** property, which includes the following:

- **1D**: This pivots around a single parameter, such as pacing, to transition between animations such as walking and running

- **2D Freeform Cartesian**: This employs two separate parameters – for instance, determining animation intensity based on a character's mood and energy levels

- **2D Freeform Directional**: Here, two parameters act as directional vectors, such as aiming a weapon in horizontal and vertical directions

- **2D Simple Directional**: Typically, this uses two parameters, often X and Y inputs, resembling a joystick's movements, to determine character direction

For nuanced hand animations using both **Grip** and **Trigger**, **2D Freeform Cartesian** is ideal. Each parameter operates autonomously: **Grip** dictates the intensity of the fist, while **Trigger** oversees the pinching gesture. Adjust **Blend Type** to **2D Freeform Cartesian**, then designate **Grip** for the X-axis and **Trigger** for the Y-axis in the **Inspector** window. In the same window, we need to define some key motions.

For a solid setup, you should consider the following four key motions depicting the extremities of the hand movements:

- **Default state**: The hand is in a default, relaxed state. This means the **Grip** and **Trigger** values are at their minimum values, which is 0. Fortunately, we don't need to create such an animation clip ourselves as the **Take 001** animation clip that comes with the Oculus Hand package we imported earlier in this chapter represents exactly this state. You can find this animation clip in the **Project** window by navigating to the `Multiplayer Scene|Oculus Hands|Animations` folder.

- **Hand pose state**: The hand pose is a pronounced pinch, where the **Grip** value is 0 and the **Trigger** value is at its maximum, which is 1. This would be the **l_hand_pinch_anim** animation clip, which is in the same folder as **Take 001**.

- **Fist pose state**: The hand is in a fist pose that is entirely clenched, with a **Grip** value of 1 and a **Trigger** value of 0. That would be the **l_hand_fist** animation clip in the same `Animations` folder as before.

- **Engaged state**: The hand is fully retracted and engaged, meaning it performs a grip and trigger movement at the same time. In this case, the **Grip** and **Trigger** values are both set to their maximum values, which is 1 for both cases. This would also be the **l_hand_fist** animation clip with both the **X** and **Y** positions set to 1.

If you did everything correctly, your **Blend Tree** should look as shown in *Figure 8.13*.

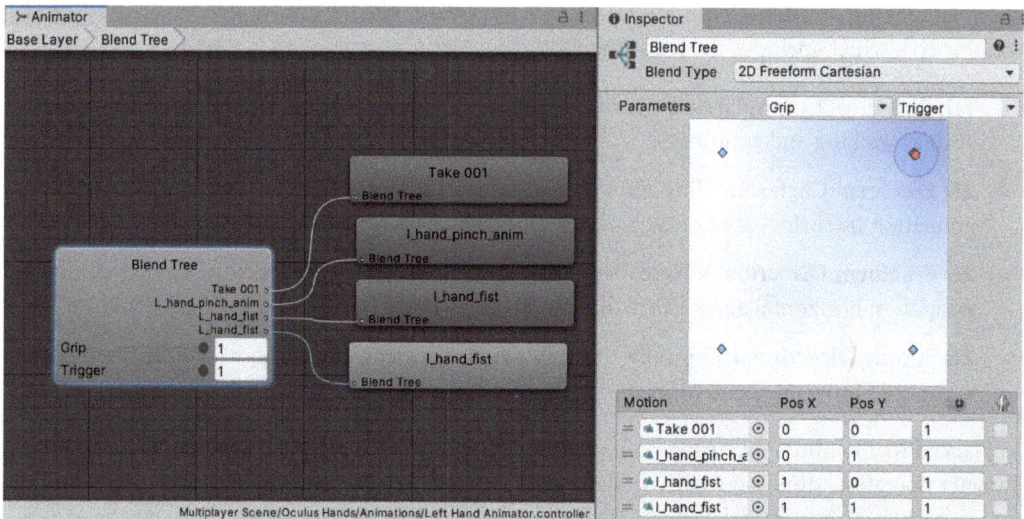

Figure 8.13 – The Blend Tree settings in Unity

Repeat this process for the **Right Hand** animator controller. Once we've done this, we can drag the just-created **Left Hand** and **Right Hand** animator controllers into the corresponding fields of the **Left Hand Model** and **Right Hand Model** prefabs.

Next, we just need to create a script that will link the **Grip** and **Trigger** variables' float values to the button inputs of our controller.

Enabling the hand animations at runtime

In this section, we'll be delving into the process of enabling hand animations in real time within our multiplayer VR application. This means that during the actual gameplay, as users interact with the virtual environment and other players, they can witness and experience hand gestures such as pinching or making a fist. This real-time interaction enhances the immersion and interactivity of our multiplayer VR game. By integrating these animations at runtime, users can seamlessly respond to game mechanics or communicate non-verbally with other players.

We are going to do this by creating a script to hold our **Hand** prefab and trigger the pinch and fist animation when the grip and trigger buttons are pressed. Now, let's begin by heading to **XR Interaction Setup** in the **Scene Hierarchy** window. Unfold it entirely to access its components. Select both **Left Controller** and **Right Controller** and proceed to add a new component via the **Inspector** window. Type HandControllerPresence into the search bar, create a new script with the same name, and then open it. We'll start by defining the necessary variables:

```
public GameObject handVisualizationPrefab;
[SerializeField] private InputActionProperty triggerAction;
[SerializeField] private InputActionProperty gripAction;
private GameObject instantiatedHandVisual;
private Animator handMotionController;
```

handVisualizationPrefab is a public GameObject variable that refers to the hand model prefab that will be animated based on the user's VR controller inputs. In our case, this will be the **Custom Left Hand** and **Custom Right Hand** prefabs that we need to drag inside these fields in the Unity Editor. triggerAction and gripAction are serialized fields, meaning they can be assigned in the Unity Editor but remain private in the script. They are designed to fetch the current values of the trigger and grip inputs, respectively.

instantiatedHandVisual and handMotionController are private variables that are used internally. The former stores the instantiated hand model in the scene, while the latter is a reference to the **Animator** component on that hand model, which is responsible for handling its animations. Let's continue with the Awake() and InitializeHandController() methods:

```
void Awake()
    {
        InitializeHandController();
    }

    void InitializeHandController()
```

```
    {
            instantiatedHandVisual = Instantiate(handVisualizationPrefab,
    transform);
            handMotionController = instantiatedHandVisual.
    GetComponent<Animator>();
    }
```

The Awake() method is called when the script instance is loaded. Here, it calls the InitializeHandController() method to set up the hand visuals and animations. This method instantiates the hand model prefab in the scene and attaches it to the object to which this script is attached. Then, it fetches and stores the **Animator** component of the instantiated hand model in the handMotionController variable. This **Animator** component will be used to control the hand's animations based on the VR controller inputs. Now, let's continue with the AdjustHandMotion() method:

```
    void AdjustHandMotion()
    {
            float triggerIntensity = triggerAction.action.
    ReadValue<float>();
            float gripIntensity = gripAction.action.ReadValue<float>();

            handMotionController.SetFloat("Trigger", triggerIntensity);
            handMotionController.SetFloat("Grip", gripIntensity);
    }
```

This method fetches the current values of the trigger and grip inputs using the triggerAction and gripAction variables, respectively. These values are in the range of 0 (not pressed) to 1 (fully pressed).

The fetched values (triggerIntensity and gripIntensity) are then passed to the hand's **Animator** component using the SetFloat() method. This effectively adjusts the hand's animation based on the real-time inputs from the VR controller. Finally, we just need to call this method once per frame to ensure that the hand's animations are updated in real time to match the user's VR controller inputs. This is done in the Update() method:

```
    void Update()
    {
            AdjustHandMotion();
    }
```

Now, we just need to assign the corresponding handVisualiationPrefab and the controller actions to the **Trigger Action** and **Grip Action** fields for both controllers, as shown in *Figure 8.14* for the left controller.

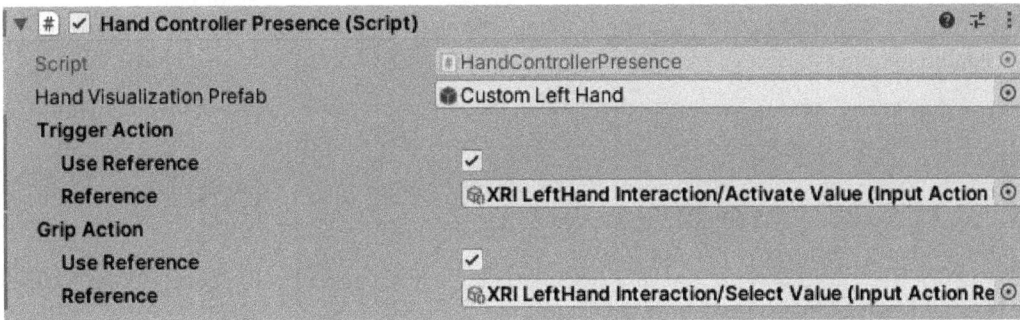

Figure 8.14 – The HandControllerPresence script and its references

You might be wondering why exactly are we using the **XRI LeftHand** references shown in *Figure 8.14*. In our scene, we are using **XR Interaction Setup**, which comes with the XR Interaction Toolkit's **Starter Assets**. This **XR Interaction Setup** uses **XRI Default Input Actions** as controller input handling.

Let's have a look at them by heading to **Samples | XR Interaction Toolkits | Your Version | Starter Assets** and double-clicking on **XRI Default Input Actions**. This will open the window shown in *Figure 8.15*.

Figure 8.15 – The XRI Default Input Actions of the XR Interaction Toolkit

The XR Interaction Toolkit action-based input system in Unity allows for device-agnostic control setups. It uses **Action Maps** to group related actions, such as **XRI LeftHand Interaction**, to handle specific actions such as **Select Value** or **Activate Value**. These actions are then tied to specific inputs, called **bindings**, such as the **Grip** button or the **Trigger** button of an XR controller, allowing for flexible and intuitive control configurations across various XR devices. You can add additional action maps, actions, and bindings or change existing ones. You can even build a complete Input Actions setup of your own under **Assets | Create | Input Actions**. However, you'll need to link this new Input Actions setup in the **Inspector** window of **Input Action Manager** under **Action Assets**. With that, we have implemented the animation of the hand models. To begin testing the application with a friend, follow these simple steps:

1. Open your Unity project and navigate to **File**, then **Build Settings**.

2. Add the currently open scene to **Build Settings**.

3. Click on the **Build** option. When prompted, choose a directory that you can easily access and remember.

4. This action will generate an executable file along with other vital scene-related files in the chosen directory. For your friend to join in, share the entire folder with them.

5. Initiate your connection to the server by pressing the **Play** button. As you await your connection, instruct your friend to launch the executable file.

And voilà! You've successfully set up your first multiplayer environment. Now, both you and your friend can interact, as depicted in *Figure 8.16*.

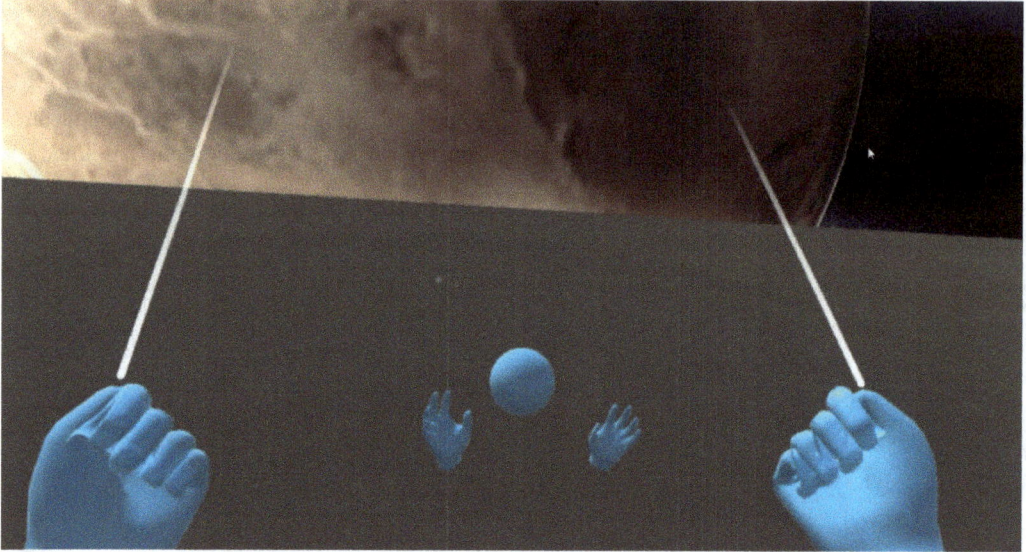

Figure 8.16 – Me and my friend in our multiplayer environment

> **Note**
>
> You might have noticed a stunning deep-space background with a picturesque planet. This aesthetic touch is provided by the Skybox Volume 2 package, which we imported from the Unity Asset Store (`https://assetstore.unity.com/packages/2d/textures-materials/sky/skybox-volume-2-nebula-3392`). It significantly enhances the ambiance of our environment.

As we approach the end of this chapter, you should now feel capable and confident in enhancing your XR experiences using advanced techniques. However, this doesn't imply that every feature, such as eye-tracking or hand-tracking, should be mindlessly applied to all your projects. Nor does it imply that all your subsequent XR endeavors must be multiplayer applications.

What it does mean is that you've broadened your horizons. You now possess a richer repertoire of interaction techniques to create your XR applications. Instead of indiscriminately applying eye-tracking to every scene element, for instance, you can selectively use it for components that would truly benefit from it. Imagine a virtual exhibition of firefighter vehicles: rather than overloading the user with information, you could seamlessly display technical specifications when the user's gaze settles on a particular vehicle. Interactive components such as animated firefighters or explorable vehicle interiors can be activated based on the user's visual focus, providing an engaging and dynamic user experience. Similarly, in a treasure-hunting game, a hint might reveal itself when a player's gaze meets a mystical mirror. The possibilities are endless!

Summary

In this chapter, you learned how to implement hand-tracking, gaze-tracking, and multiplayer capabilities into your XR scenes. Incorporating these high-level XR methods into your future XR projects will enable you to create much more intuitive, immersive, and fun experiences for users. Combined with the knowledge you gained in the previous chapters on how to create and deploy interactive XR experiences, you should feel comfortable developing a wide range of XR projects yourself, no matter if they involve coding, animations, particles, audio, or multiplayer support.

Now that you have a firm grip on the technical and programming aspects of XR, the next chapter will shift the spotlight to the business realm of XR. There, you'll explore powerful XR plugins and toolkits beyond the XR Interaction Toolkit, ARKit, ARCore, and AR Foundation that could be valuable additions to your XR development skills moving forward. The subsequent chapter won't just keep you up to date on the latest trends in XR but will also equip you with industry best practices to ensure the success of your projects.

9

Best Practices and Future Trends in XR Development

Throughout this book, you have acquired and mastered a diverse set of skills for XR development in Unity. While you should feel comfortable creating a wide array of VR, MR, and AR applications in Unity by now using the XR Interaction Toolkit, AR Foundation, ARKit, and ARCore using all the techniques you learned about in this book, we also want to help you and guide you on your XR development journey from here onward. This chapter is specifically designed for this goal. It will not only provide you with best practices in XR development that you should familiarize yourself with to become a more versatile and powerful XR developer, but also offer a comprehensive overview of XR toolkits and plugins that you might want to focus on mastering next to deepen your XR development knowledge in a particular field.

You will also learn about current trends in XR technology and applications, and we will even provide you with future trend predictions in XR made by some leading professionals and researchers in this area. By the end of this chapter, you will not only be a great XR developer, but will also feel comfortable evaluating the future path of XR technology. You will be able to better assess how you can match your XR projects to be in line with and at the forefront of these developments, as well as which skills might be interesting for you to acquire next.

This chapter is split into the following sections:

- Exploring current and future trends in XR technology and applications
- Learning the best practices for XR development
- Other useful toolkits and plugins for XR development

Exploring current and future trends in XR technology and applications

Did you know that in 2023, consumer XR headset shipments reached 17.8 million, which is over triple the number of commercial XR shipments, at 5.6 million? The business-to-consumer XR market surged to 31.1 billion US dollars, marking a 23% growth from 2022. Mobile AR dominates the XR landscape with a market value of 21.1 billion US dollars, overshadowing the VR market, which stands at 15.8 billion US dollars (`https://www.statista.com/topics/6072/extended-reality-xr`).

Being armed with the capabilities and tools to create immersive XR experiences and witnessing the rapid expansion of the XR market, you might ponder which XR niche to prioritize. This section will explore the primary domains employing XR. You'll gain insights into ongoing XR research, discover leading companies in each segment, and understand emerging trends and applications. Let's begin by examining a sector showing significant XR adoption: education.

Learning in new realities – XR in education

Imagine a world where classrooms aren't limited by four walls, and where students can traverse the universe, delve into the human body, or explore limestone caves from the other side of the globe – not through textbooks, but through immersive experiences. This isn't the realm of science fiction, but a rapidly evolving frontier in education made possible by XR, and your skills as an XR developer are crucial in bringing it to life.

An innovative technology itself is worth nothing if no one is keen on using it. At a Mexican university, more than 90% of the students involved in a research study concluded that VR aided in their understanding of 3D vectors, visualization of vector-related mathematical problems, and comprehension of angles between vectors and 3D axes (`https://repositorio.unab.cl/xmlui/handle/ria/24387`). Additionally, over 80% expressed an interest in seeking out courses with VR integration in the future and felt that getting acquainted with the VR tool was quick and straightforward.

The same also goes for teachers. A large-scale survey investigated the attitudes of over 20,000 teachers toward the use of VR for education (`https://link.springer.com/article/10.1007/s10639-022-11061-0`). This survey found that educators held a moderately favorable view of VR's role in education. The findings also revealed that as the level of VR integration increased, so did its usage frequency.

VR has been found to enhance collaborative learning by boosting engagement and motivation, facilitating remote teamwork, offering interdisciplinary collaboration spaces, cultivating social skills, and aligning with collaborative learning strategies (`https://www.frontiersin.org/articles/10.3389/frvir.2023.1159905/full`).

There's been a year-on-year increase in publications revolving around immersive VR in education over the past 20 years. This is especially true of the time period from 2017 to 2022, with an annual publication volume of 2.6 times more than the preceding period (`https://www.mdpi.com/2071-1050/15/9/7531`). These findings prove a growing research interest in using XR for educational purposes. When quantitative learning outcomes from VR applications are compared to those of less immersive teaching techniques, such as desktop computers and presentations, VR is found to provide a notable benefit in most cases (`https://link.springer.com/article/10.1007/s40692-020-00169-2`).

Several companies are leveraging the recent research findings to enhance educational experiences through VR and AR. *ClassVR* provides an immersive educational platform that pairs VR headsets with a library of curriculum-matched content. This content, known as *Avanti's World*, can also be accessed without the headsets. It's a comprehensive virtual learning environment set in a theme park format, containing six VR lands, each dedicated to various educational areas (`https://www.classvr.com/`).

Similarly, *zSpace* offers an integrated XR solution with both software and unique hardware, enhancing hands-on learning. Students, using lightweight glasses, can interact with 3D virtual models via a stylus pen. The platform offers rich content across diverse subjects, from K-8 and high-school courses in math, science, biology, and more, to vocational topics such as health science and manufacturing (`https://zspace.com/`).

Kai XR is an educational XR platform crafted specifically for teachers. This platform offers 360-degree virtual excursions, enabling students to virtually traverse global locations from their classrooms. With a commitment to child-friendly, immersive experiences, *Kai XR* offers diverse VR content, from historical sites to modern marvels. The company emphasizes its adaptability, inviting educators to either use pre-designed lessons or craft their own using *Kai XR*'s resources. The aim is to facilitate enriched learning experiences for middle-school students while honing their skills for the modern world. The platform is compatible with numerous devices, from smart TVs to VR headsets such as *Oculus Quest*.

Numerous educational XR applications have emerged on platforms such as Apple's App Store and Google's Play Store. Noteworthy examples include apps such as *Human Anatomy Atlas* and *Catchy Words*, which utilize AR, and *Ancient Egypt*, a VR experience. Additionally, VR headset manufacturers such as Meta offer a range of educational VR apps on their native platforms. This includes offerings such as *Mondly* for language learning, *Star Chart* for space exploration, and *Painting VR* for art enthusiasts.

If you are an XR developer and enthusiast, there's never been a better time to specialize in educational VR. The demand is escalating, research is encouraging, and the pedagogical implications are profound. The education sector is ripe for an XR revolution, and your expertise could shape its future.

Another driving force of XR innovation is industrial and manufacturing use cases, about which you will learn more in the upcoming section.

XR in industrial settings and smart manufacturing

Almost 13% of entrepreneurs in the US are currently integrating XR solutions into their companies. This percentage underscores a growing interest in and acceptance of the technology. Moreover, projections indicate that close to 23% of companies plan to incorporate these technologies within the next three years. This provides a substantial window of opportunity for XR developers to capitalize on a market that's set to expand.

One of the significant benefits of integrating XR technologies into the industrial sector is the potential for a drastic reduction in workplace accidents – by up to 70%. By identifying potential hazards and dangerous maneuvers using XR, industries can create a safer working environment. This focus on safety and efficiency is likely to drive demand for XR solutions that can deliver such results.

Data captured from the US market suggests that around 2.5% of companies are fully utilizing XR technologies. These modern entities, often referred to as *smart factories of the future*, have a holistic technological ecosystem that's conducive to XR solutions. Tapping into this segment allows developers to push the boundaries of what's possible with XR, setting benchmarks for the industry (`https://doi.org/10.1016/j.procs.2022.09.357`).

In the realm of smart manufacturing, AR in particular is making strides. By using AR devices such as smartphones or headsets, workers receive real-time information and guidance while executing tasks. For instance, an AR headset can display step-by-step assembly instructions and provide feedback if an error occurs or if a tool is nearing the end of its usability. Another useful application of AR in manufacturing is one where a customer, requiring assistance with repairs, is guided remotely by an expert using AR glasses, such as Microsoft's *HoloLens*.

As the usage of AR increases workers' accuracy and efficiency, adopting this technology in a smart manufacturing context leads to overall increased productivity in manufacturing facilities. For onboarding and training of new staff, AR can visually represent intricate tasks and processes, ensuring better understanding and retention. Managers can quickly ascertain the production status via their mobile devices as they navigate the factory floor (`https://www.mdpi.com/2079-9292/12/7/1719`).

In recent literature, assembly has been found to be a leader in AR adoption in manufacturing. This prominence stems from assembly activities being visual-centric, demanding significant operator involvement. Comparative studies have shown AR improves efficiency in maintenance and assembly, shortens maintenance time, and enhances task quality over conventional tools.

Quality assurance, too, has benefited from AR. Initially using AR for projecting 2D information, it now uses spatial AR for quality checks, as with welding spot inspections. Spatial AR, combined with real-time 3D metrology data, aids in assessing parts' surface quality, ensuring precise welding, and supporting packaging sectors such as die-cutter setups, reducing errors and costs.

Recent studies have also utilized AR for quality control on automotive parts, eliminating the need for operators to constantly refer to static documentation, streamlining processes, and increasing efficiency. Tests have revealed that AR systems reduced execution time for complex quality control procedures by 36%. Other applications, based on a user-centered design, have showcased AR's potential in assisting workers with design discrepancy detection. The AR tool's efficiency is evidenced by the minimized cognitive load it places on users, as they must no longer split their attention between design data and physical prototypes.

Another study in the automotive domain developed an AR system to correct alignment errors during car body fitting, significantly expediting the process. The AR system provided real-time guidance, making the procedure almost four times quicker, with more consistent results regardless of operator experience. Future enhancements aim to further minimize setup times and implement artificial neural networks for error detection (`https://www.mdpi.com/2076-3417/12/4/1961`).

In industrial AR settings, tracking techniques play a pivotal role in accurately determining the position and orientation of objects and devices, ensuring an immersive and interactive AR experience. Based on the provided information, the most used tracking techniques in these industrial scenarios are the following:

- **Marker-based tracking:** Among the most prevalent tracking methods in the manufacturing context, marker-based tracking holds a dominant position. We discussed this technique in great detail in the *What are marker-based and markerless AR?* section of *Chapter 4*. The primary advantages of marker-based tracking for industrial use cases are its speed, simplicity, and robustness. It has been extensively applied in scenarios such as facilitating AR instructional systems for assembly processes, assisting maintenance processes, guiding shop-floor operators, and supporting welding processes in the automotive industry. The method's wide application can be attested by the fact that out of 200 selected research articles related to AR usage in an industrial setting, 63 articles (31.5%) focused on marker-based tracking, indicating its significance in AR-based manufacturing applications.

- **Markerless tracking**: Markerless tracking, another method that we discussed extensively in the *What are marker-based and markerless AR?* section of *Chapter 4*, takes the second spot in terms of the dominance of frequently used AR tracking techniques in industrial settings. Examples of its application include using 3D point clouds of workstations for quality control procedures in the automotive industry, or integrating cloud databases for fault diagnosis.

- **Hybrid tracking**: An emergent trend, hybrid tracking combines the strengths of computer vision-based techniques, such as marker-based or markerless methods, with sensor-based approaches. By doing so, it achieves faster tracking speeds, reduces latency, and alleviates the computational burden inherent in markerless tracking algorithms. Incorporating additional sensor data can also enhance the performance of other tracking methods.

- **Model-based tracking**: Used when a 3D CAD model of the tracked object or part is available, this method analyzes and recognizes the pose and position of objects based on these models. Even though its popularity has been on the rise due to its integration into AR software development platforms such as Unity and *Vuforia*, it's still less dominant than the other methods, with 13 out of 200 research articles (6.5%) implementing this technique.

- **Sensor-based tracking**: This method is relatively less favored in AR solutions for manufacturing, primarily because of the complexities and costs associated with deploying sensors, especially in indoor environments. Challenges include potential blockages of sensor signals by equipment, machines, or other objects, making it less effective (`https://www.mdpi.com/2076-3417/12/4/1961`).

While various tracking techniques cater to different AR applications in manufacturing, marker-based tracking remains the most favored due to its simplicity and robustness. However, as industrial settings evolve and technology continues to advance, we're witnessing a gradual shift toward more complex yet versatile tracking techniques such as markerless and hybrid methods. If you're interested in exploring the world of industrial XR, getting to know the pros and cons of these tracking techniques first-hand should be one of your main priorities moving forward.

A rising star in the industrial XR space called *RealWear* specializes in hands-free AR wearable solutions. Its AR wearable devices allow workers in fields such as energy and manufacturing to access crucial information while keeping their hands free for tasks (`https://www.realwear.com/`).

TeamViewer is also betting on AR for industrial use cases. Its AR platform *TeamViewer Frontline* is designed to usher the deskless industrial workforce into the digital age. Leveraging state-of-the-art wearable technology, *Frontline* streamlines manual processes across industries, bridging the gap between digitally proficient office workers and hands-on industrial teams. Its suite of XR solutions offers seamless integration, extensive hardware compatibility, and user-friendly interfaces (`https://www.teamviewer.com/en/products/frontline/`).

Similarly, *PTC's* Vuforia offers a comprehensive suite of AR solutions tailored for industrial use cases. From enabling service technicians to visualize and get insights about machinery using AR to assisting in design reviews and factory layout, Vuforia's AR platform caters to a vast array of industrial needs.

As you can see, the industrial sector presents a fertile ground for XR developers that's ripe with opportunities for innovation, disruption, and growth. The tangible benefits of XR, coupled with the trends in adoption and the challenges faced by industries, make it a lucrative domain for XR developers to focus their efforts on.

Another sector that benefits hugely from the integration of XR is healthcare.

Bridging the gap from real bodies to virtual worlds – XR in healthcare

VR has been garnering attention for its potential in medical education. A comprehensive meta-analysis, spanning studies from 1990 to 2019, aimed to gauge VR's efficacy in anatomy teaching. The study's findings revealed that VR adoption in anatomy instruction improves students' test results compared to traditional methods. The authors' analysis underscores the potential of VR as a valuable tool in elevating students' grasp of anatomy (`https://bmcmededuc.biomedcentral.com/articles/10.1186/s12909-020-1994-z`).

UbiSim is an interesting company in this domain. It provides nursing students with a fully virtual simulation lab, immersing them in environments where they engage with realistic patients. This tool aims to refine their patient engagement, clinical reasoning, and decision-making abilities, offering varied scenarios from emergency responses to childcare (`https://www.ubisimvr.com/`). Other companies offering similar platforms for surgical VR trainings are *PeriopSim VR*, *Osso VR*, and *Oxford Medical Simulation*. The latter delivers comprehensive VR training for both the medical and nursing sectors. The platform's content is available on regular computers or VR headsets. Emphasizing learner-centric methods, it offers realistic patient care scenarios and teamwork simulations. Several esteemed institutions such as Oxford, Manchester, and Edinburgh Universities have incorporated this platform into their curricula (`https://careers.oxfordmedicalsimulation.com/`).

Another big leap in XR technologies for healthcare is the use of VR for rehabilitation purposes. There's been a surge in the number of individuals requiring rehabilitation, especially among older individuals and those with disabilities, chronic diseases, and functional and cognitive impairments. These individuals face challenges in movement, sensation, balance, and cognition, which significantly impact their quality of life, careers, and social engagements.

Traditional rehabilitation encounters issues such as stationary rehabilitation centers, resource scarcity, monotony in the training process, soaring treatment expenses, and a lack of self-driven encouragement and automatic guidance. This often leads to decreased confidence and diminished outcomes in the rehabilitation journey.

With VR, the rehabilitation process is being completely transformed. VR provides an immersive experience, motivating patients and enhancing their active involvement. This not only overcomes the stationary nature of traditional centers but also fills the gap of lacking resources. Moreover, VR-based systems paired with appropriate sensors can closely monitor and record patient movements and biological data, offering an avenue for refining and personalizing rehabilitation programs.

Contributions to this research domain stem from 63 countries and 1,921 institutes, highlighting the global attention and collaboration that VR rehabilitation is garnering. Research trends in VR rehabilitation are areas such as kinematics, neurorehabilitation, brain injury, exergames, aging, motor rehabilitation, and cerebral palsy, among others. This diversity underscores the broad applications and scope of VR in various rehabilitation sectors (`https://www.ncbi.nlm.nih.gov/pmc/articles/PMC10028519/`).

An interesting company in this space that specializes in offering a wide range of different rehabilitation programs in VR, from chronic back pain to Parkinson's disease, is *Penumbra* (`https://www.realsystem.com/`).

VR development for surgical trainings, rehabilitation, or other fields of healthcare is not just an exciting intersection of healthcare and technology, but also an area that's ripe for innovation, collaboration, and immense growth potential. Given the evident gaps and burgeoning demand, there's a compelling reason for XR enthusiasts and developers like you to specialize in this field, driving the next wave of meaningful XR applications.

The idea of enriching the gaming industry with XR applications may not initially seem as profound as developing VR trainings for surgical procedures. However, as you'll discover in the following section, XR holds the potential to enable entirely novel gaming experiences, thereby influencing this multi-billion-dollar industry in ways that we cannot fully predict at this moment.

Gaming across realities – XR trends in gaming

Gaming has a historical symbiosis with XR; it is in the rich and diverse gaming landscapes that XR found its early applications and extensive testing grounds. The industry has continually pushed the boundaries, seeking to offer increasingly immersive, interactive, and realistic experiences, which naturally positioned it at the forefront of XR adoption and innovation.

Top-tier companies and gaming studios have been pivotal in nurturing this space. Renowned for its graphics processing units, *NVIDIA* is pivotal in the XR world. Its *RTX* series GPUs, designed with ray tracing capabilities, offer enhanced realism in VR environments. It has also worked on AI-driven advancements that improve XR experiences, such as *deep learning super sampling*.

Sony introduced the *PlayStation VR*, a VR headset that integrates with its PlayStation gaming consoles. Games such as *Astro Bot Rescue Mission* and *Blood & Truth* offer immersive VR experiences exclusive to the platform.

Renowned game studios such as *Ubisoft*, *EA Sports*, and *Epic Games* have also been proactive in infusing XR into their gaming ecosystems, thereby transforming the way we game today. They've partnered with other tech leaders to create XR-supported games that offer unprecedented levels of immersion and interactivity.

Apart from these tech and gaming giants, several start-ups and smaller entities are throwing their hats into the ring, eager to carve niches for themselves. Companies such as *Magic Leap* and *Niantic* (the name behind *Pokémon GO*) have been working tirelessly to build ARs that merge the physical and digital worlds in gaming scenarios. Other companies to watch include *Resolution Games*, a studio focusing on VR games and publisher of well-known VR games such as *Demeo*, and *Rec Room Inc.*, known for its VR social platform that integrates gaming elements.

For XR developers looking to immerse themselves in an arena rife with opportunity, the gaming industry presents a landscape fertile with potential. The industry offers a rich variety of genres and narratives to explore and innovate upon, from fantasy worlds to simulations grounded in reality.

Furthermore, with gaming being a multi-billion-dollar industry, it promises not only creative satisfaction but also substantial financial rewards. Developers have the chance to work on cutting-edge technology, pushing the boundaries of what is possible and shaping the future of entertainment.

Moreover, the gaming audience, traditionally receptive to new technologies and innovations, provides a ready market that is eager to embrace the novelties that XR can bring to the gaming space. This receptive audience can offer a particularly straightforward path to commercial success for well-conceived XR gaming products.

A research study explored the factors that influence people's acceptance and use of VR games through an online survey involving 473 VR gamers. The researchers investigated the impact of hedonic and utilitarian aspects on VR gaming. Hedonic aspects refer to the pleasure and enjoyment derived from playing VR games. It encompasses aspects such as fun, enjoyment, and the immersive experience that these games offer. Utilitarian aspects relate to the practical benefits that can be derived from playing VR games, such as health and well-being improvements that some games can facilitate. The findings of this study showed that the hedonic aspects were a stronger driver for people playing VR games compared to the utilitarian aspects. Despite potential inconveniences such as physical discomfort and VR sickness, the researchers found that these factors did not significantly reduce the intention to use VR games or the level of immersion experienced by the players. The ease of using VR systems was found to be crucial in encouraging both the pleasure-driven (hedonic) and practical (utilitarian) usage of VR games (`https://link.springer.com/article/10.1007/s10055-023-00749-4`).

The results provide XR developers with important insights; understanding that players prioritize enjoyable and immersive experiences over practical benefits could guide the design process. Additionally, ensuring the system is user friendly can foster both enjoyment and practical utility. Thus, XR developers aiming to specialize in gaming should focus on creating immersive, fun, and easy-to-use experiences, while not being overly concerned with the potential physical discomfort and VR sickness that might be associated with VR gaming.

Another study aimed to understand player complaints about PC-based VR games by carrying out an empirical study of 750 PC-based VR games and 17,635 user reviews on the gaming platform *Steam*. It found that most newly released VR games support multiple headset categories and play areas. There has been an increase in support for smaller-scale play areas over time. Even though the median price for VR games has doubled, complaints about games being overpriced for their content have sharply decreased. Similarly, complaints about cybersickness are low and have been declining steadily over time. From 2018 onward, most of the complaints have shifted from VR comfort issues to game-specific problems (`https://ieeexplore.ieee.org/document/9347709/`).

The findings suggest that the PC-based VR gaming market is becoming more sophisticated. Concerns traditionally centered on VR comfort, such as cybersickness, have lessened in prevalence. This means XR developers specializing in gaming should now prioritize enhancing game design and gameplay aspects, while ensuring VR comfort remains at its current standard or better.

A study using data from 515 Pokémon GO players found that nostalgic feelings about the Pokémon franchise fueled the players' imagination of AR game content in the real world. This, in turn, nurtured affection toward the AR content, enhancing the sense of meaningfulness derived from playing. Similarly, social involvement, including community identification and social self-efficacy, elevated the meaningfulness of the gaming experience (`https://www.sciencedirect.com/science/article/pii/S0747563221001394#sec7`).

Researchers from another study developed *ARQuiz*, a game for public exhibition attendees of a Finnish science center. The ARQuiz game enabled visitors to engage in a virtual quiz related to the exhibits. During the exhibition visit, players used their smartphones to locate AR hints for answers to the quiz questions related to the nearby exhibits. In addition to the main quiz gameplay, users could interact socially by leaving messages for other players. Their study found that the AR application positively influenced the overall visitor experience at the public exhibition. Visitors who enjoyed ARQuiz also enjoyed the exhibition more, performed better in the quiz, and felt more sociable after the exhibition. ARQuiz was perceived as enjoyable and offered a different user experience compared to traditional exhibitions (`https://ieeexplore.ieee.org/document/8861040`).

The gaming industry emerges as yet another compelling avenue for XR developers. Regardless of the specific industry you intend to target with XR applications, there exist certain essential best practices and design principles that your XR applications should consistently follow. These will be presented in the upcoming section.

Learning the best practices for XR development

This section will provide you with best practices and design patterns for all your XR projects, regardless of the industry you are creating them for. With this knowledge, you will be able to make the right hardware, software, and design decisions for each of your XR projects. Let's start by exploring the hardware considerations you should keep in mind at the beginning of your XR projects.

Hardware considerations in XR development

In the realm of XR, hardware is more than just a vessel for experiences; it's a determinant of their quality. The most powerful VR device isn't universally ideal for every XR application, much like how buying a supercomputer would be overkill for the sake of creating a PowerPoint presentation. Consider a fair scenario: would you prefer a lengthy VR setup with external sensors, demanding PC requirements, and full-body tracking to show off a VR representation of a solar panel? Or would standalone VR devices such as those from Meta or PICO, known for their user-friendly operation, be more adequate for this task?

The right hardware strikes a harmony of performance, adaptability, and ease of use. Some pivotal factors to consider are the device type, performance, and testing. Let's have a closer look at different device types and their strengths in the next section.

XR device types and their strengths

Understanding the constraints of XR devices is essential for XR developers aiming to craft immersive and smooth experiences. Each XR device, whether it's a smartphone, AR glasses, or a dedicated VR headset, is tailored for different scenarios based on its computational power.

For instance, a smartphone, with its versatility and widespread use, is ideal for AR applications such as Pokémon GO or AR-driven shopping experiences where the user can visualize products in their real environment. AR glasses, such as Microsoft's HoloLens, are more suitable for specialized tasks such as hands-free work instructions or architectural visualizations. In contrast, dedicated VR headsets, such as the Oculus Quest or the *HTC Vive*, are designed for deep, immersive experiences where the user is transported into an entirely virtual world. The following are some of the most common use cases that suit the different types of XR hardware:

- VR headsets:

  - Fully immersive gaming experiences with a 360-degree field of view

  - Simulation training where users can practice tasks such as surgeries or navigating airplanes in safe virtual spaces

  - Experiencing places and historic sites without leaving home

  - Architectural visualization to walk through your future home before it's built

  - Virtual offices where global teams can collaborate in real time with avatars

- AR glasses:

  - Step-by-step overlay guidance for complex tasks

  - Real-time directions overlay on the real world, combined in an industrial setting where users should move hands-free

  - Surgeons receiving overlays of important info while operating without the need to move their heads constantly between the patient and a screen

  - Technicians getting schematics or guidance overlaid on machinery they're fixing

- Smartphones (AR):

  - Gaming experiences such as Pokémon GO where players interact with real-world locations

  - Previewing furniture or clothing items in your space before buying

  - AR-driven walking or driving directions such as the ones offered by Google Maps

  - Providing tourists with historical info on landmarks or interactive learning modules

  - Augmenting video calls or video content with digital enhancements

  - Real-time guidance on tasks such as home repairs

Once you have chosen the right hardware for your XR project, you will need to be aware of the performance challenges that are common in this field. Let's go through them in the upcoming section.

Performance challenges of XR devices

Creating XR applications that depict detailed 3D environments, such as architectural visualizations in VR, requires addressing performance challenges. Older devices or entry-level VR headsets with limited computational capabilities may not render these details smoothly, resulting in lags or visual disturbances. Similarly, an AR application loaded with high-resolution textures and complex 3D models may run effortlessly on advanced devices such as the latest iPhone, but might falter on older ones due to memory constraints.

The preliminary step in creating an XR application is to assess the 3D models you're incorporating. Unity, for example, provides information on the polygon or triangle count of an imported 3D model. Typically, a 3D model's polygon often equates to a triangle consisting of three vertices. Thus, a model's intricacy directly relates to its polygon count, which becomes pivotal for devices with limited processing capabilities such as standalone VR headsets or smartphones.

Referencing documentation provides an understanding of device-specific polygon count limitations. For example, the Oculus Quest 1 documentation suggests a triangle count cap of 350,000-500,000, while Quest 2 can handle 750,000 to 1 million polygons (`https://developer.oculus.com/documentation/unity/unity-perf/`). Note that this pertains to the cumulative polygons rendered in a scene, not just an individual model. Hence, for optimal performance, the aggregate polygon count of all entities in a scene shouldn't surpass this figure, though exceptions exist based on other optimized factors such as shaders or physics.

Suppose a client wants to showcase five high-resolution car prototypes in VR using the standalone mode of the Quest 2 headset. However, each car model has 1 million polygons, so the total number of polygons for the five models would be 5 million, which exceeds Quest 2's recommended limit of 2 million polygons. There are a few things that developers can do to address this issue:

- **Use PC VR mode**: Opt to run Quest 2 in PC VR mode paired with a robust computer, sidestepping the need for optimizations. The reasoning behind this is that PCs possess greater computational prowess, eliminating the typical hardware constraints of standalone VR headsets.

- **Utilize mesh decimation**: Mesh decimation is a powerful technique to significantly diminish a model's polygon count without notably downgrading its visual appeal. Implementing this method requires proficiency in 3D modeling software and an understanding of the underlying geometry of the models being simplified. There's no hard limit, but the key is to strike a balance: reducing polygons effectively while ensuring the model retains its essential characteristics and aesthetic value.

  To learn more about mesh decimation, please refer to this insightful blog post: `https://odgy.medium.com/mesh-decimation-done-right-95245c4b5f52`

- **Optimization techniques**: Although standalone VR headsets can accommodate scenes exceeding their recommended polygon count, achieving this demands numerous optimizations. Adopting straightforward shaders, implementing baked lighting, and opting for reduced texture resolution are a few methods of doing so. However, from our experience, it's advisable to stay within a million polygons over the stipulated limit for such optimizations.

Lastly, you should continuously test and assess your XR application throughout its entire development process to make sure it is in line with your hardware's capabilities. Tools such as Unity's **Profiler** serve as valuable aids. In the next section, you will learn how you can use it.

Testing and profiling the application

Unity's Profiler is an invaluable tool for diagnosing performance issues and optimizing your VR experiences. The Profiler provides real-time data about your application, capturing details on CPU, GPU, memory usage, rendering, physics, and more. Implementing and using the Profiler within a VR experience is similar to using it for any other Unity project, with some VR-specific considerations. The following is a step-by-step guide on how to implement Unity's Profiler for your VR experience:

1. Open your project in Unity. Go to **Window** | **Analysis** | **Profiler**, or simply press *Ctrl/Cmd + 7* to open the **Profiler** window. If you are using a standalone headset, ensure that both your PC and the headset are on the same network.

2. If you have a PC-based VR headset, simply click the **Play** button in Unity to start your VR experience. While your VR experience is running, the **Profiler** window will update in real time, showing data captured from your application. For standalone VR headsets, you must first build and deploy your application to your VR device.

3. In Unity's **Profiler** window, click on the **Active Profiler** dropdown. You should see your VR device listed there (given it's on the same network). If not, ensure Unity and your device can communicate over your network. Select your VR device from the list. Unity will start profiling directly from the device, and the **Profiler** window will update in real time.

While many of the metrics in the Profiler are applicable to all Unity applications, with VR, you should pay special attention to the following:

-   **Frame time**: Ensure you're consistently hitting the frame rate required for your VR headset (e.g., 90 Hz for Oculus Rift, 72 Hz for Oculus Quest). Missing frame rates can lead to motion sickness in VR. You will find the recommended frame rate in the documentation of your VR hardware.

-   **Render thread**: VR requires rendering two views (one for each eye), which can be more demanding. Look for any spikes or bottlenecks here.

-   **GPU**: High-quality shaders and post-processing effects can be especially demanding in VR due to the need to render them twice (for each eye). Always test their impact in VR.

Once you've identified areas where your VR experience might be encountering performance hitches, you can dive deeper into the specific sections (such as **CPU**, **GPU**, and **Memory**) in the Profiler to get a detailed performance breakdown. Look for any unexpected spikes or prolonged high usage. These could be indicators of potential issues such as inefficient scripts, too many active physics objects, or overly detailed models. Make adjustments in your project based on your findings. For instance, simplify complex meshes, optimize shaders, and reduce the number of active physics objects. After making changes, test your VR experience again to see whether the issue has been resolved and to ensure no new issues have cropped up.

> **Tips**
>
> These three important aspects are often forgotten when profiling XR applications in Unity:
>
> 1. Unity has a *Deep Profiling* mode that provides more detailed information, but it comes at a performance cost. Use it when you need an in-depth analysis.
>
> 2. Remember to turn off any debug logs in the final build as they can impact performance.
>
> 3. For VR, always strive for consistent and smooth performance. Even minor hitches can be disorienting and uncomfortable for users.

By using the Profiler regularly throughout your VR development process, you can ensure that your applications run smoothly and provide the best possible experience for your users. However, a smooth-running experience does not guarantee that users will enjoy your app. Let's have a closer look at the most important aspect to focus on before designing your app: your audience.

Understanding your audience

One of the foundational steps in crafting a successful XR application is understanding who will be using it. Just as a carpenter wouldn't start building without a blueprint, a developer shouldn't start creating without a clear picture of their audience. This section will guide you through the process of identifying and understanding your target users.

Before you dive into the development of your VR application, you need to ask: "Who is this for?" The answer will influence almost every other decision you make. For instance, a VR application for children might prioritize colorful graphics and simplicity, whereas one for professionals in a given industry might focus on functionality and data visualization. Consider age, gender, education, and occupation. A VR game targeted at teenagers will have different design elements and challenges than one aimed at adults.

Another important question to answer in this context is whether your audience is familiar with VR. If not, your application might need to incorporate introductory tutorials or more intuitive controls. A tech-savvy user might crave advanced features and customizations, while others might want a more streamlined, plug-and-play experience. By identifying these user needs, you can ensure that your application will be relevant, user friendly, and appealing to your audience.

It's always wise to assess the landscape before planting your flag. Look at the VR applications currently available that serve a similar purpose or target audience to yours. Identify what's missing in current offerings. Is there a feature or an experience that users are asking for in the reviews but isn't yet available? Staying with user reviews, the insights you will receive just by reading through reviews of similar applications are invaluable. Users will often detail what they love, what they hate, and what they wish was included.

Creating user personas is another common method in product design, and its importance is magnified in the world of VR, where experiences are so personal and immersive. A user persona is a fictional character that represents a segment of your target audience. They come with a name, a background, preferences, and challenges. Personas help make abstract user needs more tangible. When making decisions, you can ask: "Would this feature benefit Anna, the tech-savvy college student who loves gaming?" or "Would this interface be intuitive for Raj, the middle-aged architect who's new to VR?" Incorporating user personas into your planning phase can be the difference between a VR application that feels generic and one that feels tailor-made for the user.

Understanding your audience isn't just the first step in VR application development; it's a compass that should guide every subsequent decision. Making the effort to truly understand your users ensures that the final product will not only meet but exceed their expectations. After you've understood your audience, you must also ensure that you can cater to its needs. No matter whether you're working alone or in a team, efficient project management is crucial to completing your XR application in a reasonable time and manner. Let's dive deeper into this topic in the next section.

Efficient project management for XR

Every great creation begins with an intention, a clear vision of what one hopes to achieve. Setting tangible goals ensures that the development process remains focused and the end product aligns with the envisioned purpose. This section sheds light on how to set effective goals, prioritize features, and determine the criteria for success.

Just as a sailor wouldn't set out without a destination in mind, developers should have a clear objective for their VR application. Begin by asking: "Why are we building this?" Are you looking to entertain, educate, train, or provide a unique service? Pinpointing the purpose will shape the app's design, content, and user experience. Once you can answer this question, determine the boundaries of your application. If it's an educational VR app, will it cover one subject or offer a broad curriculum? Setting the scope ensures your efforts remain targeted and manageable. In this context, you should also specify what the target user hopes to achieve with your application. Whether it's mastering a new skill, being entertained, or solving a specific problem, understanding this can help refine your objectives.

With all of these questions settled and an array of possible features to integrate, you need a strategy to prioritize. The following are some important considerations you should make:

- List out all the features you envision for your application. Then, categorize them into essential functions and those that are additional enhancements. This helps in focusing on what's crucial first.

- Use your user personas or early feedback to determine which features are most desired by your audience.

- Be realistic about time, budget, and technical constraints. Some features might be fantastic on paper but could be resource-intensive to implement. Keep in mind the number of people and the competencies within your team.

Defining success is as crucial as setting the initial goals. How will you know whether your VR application meets, or hopefully exceeds, expectations? For this, we recommend you establish success metrics right at the start of your XR endeavor that you can evaluate post-launch, as well as during user testing. Track metrics such as the average session duration, frequency of use, and user progression within the application to gauge engagement levels. Likewise, monitor crash rates, glitches, and other performance-related issues. A technically sound application is key to retaining users. Use your metrics not just as a report card but as a roadmap for future updates, refinements, and even potential expansions.

For your XR endeavors to succeed, you should also be familiar with some essential techniques and tools for managing your projects in XR. Let's go through them step by step:

- **Mind mapping and brainstorming sessions**: These enable you to harness collective intelligence, propelling the clarity and creativity of your concept. For these sessions, assemble a diverse group of people, combining technical developers, designers, and even potential end users. You should do this even if you are a solo developer. Their combined insights can lead to enriched ideas. For these sessions, we recommend you use tools such as *Miro* or *Lucidchart*.

- **Game design documentation (GDD)**: A GDD acts as a project's blueprint, weaving together the narrative, mechanics, interactions, and level designs. Rather than just sketching a user's journey, as with traditional storyboarding, a GDD delves deeper. It outlines the intricacies of user interactions in the XR environment, lists mechanics, sets the governing rules, maps out the challenges within each level, and ensures a seamless user experience by detailing menus. This holistic approach streamlines development and sets the stage for a successful project.

- **Creating 3D wireframes**: While traditional wireframing tools assist in plotting out digital interfaces, XR demands a more 3D approach. Tools such as *Sketchbox* and *Gravity Sketch* let developers and designers craft 3D wireframes, allowing a spatial understanding of the XR environment.

- **Managing your code base**: Once the project is planned, it needs to be managed. In particular, managing code base changes in complex XR projects can be a daunting task. **Git**, a distributed version control system, will become your daily companion in such scenarios. Git allows developers to work simultaneously on different features or bugs without affecting the main code base. Once a feature or fix is ready, it can be merged back into the main branch. Every change, every merge, and every version is tracked. This means if something goes wrong, you can revert to a previous state of your application easily. Git provides a platform where multiple developers can contribute without stepping on each other's toes. Each contributor can work on their local copy and then push changes to a shared repository.

  Both *GitLab* and *GitHub* offer platforms that extend the capabilities of Git, making the collaborative process even smoother. Beyond just version control, these platforms offer a suite of tools such as code reviews, continuous integration, and continuous deployment, which streamline the development process. You can also report, track, and assign bugs or features with them. This keeps the team aligned on priorities and ensures that issues are resolved methodically. Both tools also come with **kanban boards**. A kanban board is a visual tool to manage tasks and workflows. As tasks move from one stage to another, such as from **To Do** to **In Progress** or **Completed**, they can be dragged across columns on the board. This provides a quick overview of project progress and bottlenecks and is perfectly suited for agile projects.

Planning an XR application is a multifaceted process, blending traditional techniques with XR-specific tools. This integration ensures that the foundation of the XR application is robust, clear, and primed for successful development. Whether you're sketching out the user's journey or building a 3D prototype, each planning stage is a step closer to realizing a compelling and impactful XR experience.

In the next section, we are focusing on best practices that ensure a good user experience.

Ensuring a good user experience

Ensuring a good user experience in XR applications goes beyond just graphics and performance; it delves deep into intuitive design and user interaction. From grasping the basics of XR UI/UX design to creating inclusive interfaces and understanding the importance of feedback, every aspect plays a pivotal role in defining a seamless and immersive experience for the user.

Traditional UI/UX, whether for web or mobile apps, focuses on the spatial relation on a flat screen. Designers ask: "How does one element look next to another?" In XR, the questions evolve. The spatial relation is not just side-to-side, but also in-depth. It's about how one element exists in relation to another within a 3D space.

XR environments also introduce challenges not found in conventional design. For instance, how do you ensure that a user can comfortably read text in VR? Or in AR, how do you design an interface that complements, rather than clashes with, the real world?

As XR has matured, certain patterns and guidelines have emerged that aid in creating intuitive and immersive experiences. Just as a hamburger icon signifies a menu in traditional apps, certain gestures or symbols are becoming standardized in XR. For instance, a hand pinch in VR might signify selection, while a swipe in AR might rotate an object. Physical feedback, such as a controller's vibration in response to an action, can significantly reinforce a user's actions within the XR environment. No matter what kind of XR application you are developing, a good rule of thumb is the following: leverage familiar actions from the real world (such as grabbing or throwing) to make virtual interactions in your XR application instantly understandable.

Similarly, XR interfaces need to be designed with user comfort at the forefront. Elements should be within a comfortable reach and view, ensuring users don't need to make awkward movements or strain their eyes. This also involves considering ergonomics, ensuring that interactions don't lead to physical strain over extended periods. There's a temptation to include a plethora of interactions and details in XR applications, but this can often lead to confusion. Clarity and simplicity should be the watchwords. If an interface element or interaction doesn't serve a clear purpose, it's worth reconsidering its inclusion.

With the digital revolution, one would assume that accessibility barriers have been significantly reduced. However, without careful design, XR can inadvertently introduce new ones. It's essential that XR applications are not just for the many but for all. There are three important things to keep in mind for any of your future XR applications:

- Not all users interact with XR in the same way. Some might have visual impairments, others could have auditory challenges, and yet others might face mobility restrictions. For instance, subtitled audio can aid those with hearing impairments, while voice commands can help those with mobility challenges.

- One-size-fits-all rarely works in design, especially in XR. Providing users with the ability to adjust settings to their comfort can make a massive difference. This includes scaling the UI for those who might find default sizes hard to read or adjusting brightness for those sensitive to intense light. Audio levels, especially in VR, can be adjusted to ensure users can comfortably immerse themselves in the environment.

- Inclusivity in XR is not just about adding features; it's about adopting a mindset. When the design process starts with considering all potential users, regardless of their challenges, it leads to a product that's not only more robust but also more universally enjoyable.

The potential of XR is boundless. However, this potential can only be realized when designers place users at the heart of the design process. By focusing on intuitive interactions and championing inclusivity, XR applications can offer transformative experiences for everyone.

Another important aspect to ensure a good user experience is the gathering of user feedback. The development of any XR application is only as good as the feedback it receives. In XR, where users are navigating complex 3D environments and engaging in novel interactions, understanding their experiences is crucial. There are a couple of possible ways in which to collect user feedback:

- **Playtesting**: This is one of the most direct ways to gather insights. By observing users as they navigate through an XR experience, you can identify pain points, moments of confusion, or elements that work exceptionally well. It provides a firsthand look into how users engage with your application, making it invaluable for XR development.

- **Surveys and questionnaires**: These can be used post-experience to gather structured feedback. They can focus on specific aspects of the XR application, such as the intuitiveness of interactions, visual aesthetics, or overall user satisfaction. Tailored questions can extract deep insights into particular elements, while open-ended questions can provide unexpected but valuable perspectives.

- **Spontaneous feedback**: Beyond structured playtests, sometimes just allowing users to interact with an XR application and share their thoughts spontaneously can yield the most candid feedback. These sessions can be more free-form, letting users explore at their own pace, with developers noting reactions, comments, and behaviors.

Not all feedback will be actionable, and some might even be contradictory. The key is to look for patterns and consistent pain points or praises across users. Tools such as heat maps in combination with eye-tracking, to visually represent where users look or interact the most, can be invaluable. Likewise, feedback about physical discomfort, confusion, or moments of delight should be given priority.

Based on such analyses, your XR experience should undergo adjustments. This could mean redesigning certain UI elements, optimizing performance, or even introducing new features. The goal is to enhance what works and fix or eliminate what doesn't.

To avoid criticism on interactions not feeling natural or intuitive, you should focus on choosing common input configurations from the beginning. Let's discuss them in the next section.

Input devices

In the realm of XR, the mode of interaction is a pivotal aspect that can significantly influence user immersion and engagement. Modern XR experiences offer a spectrum of interaction modalities, from controllers to direct hand-tracking and gesture recognition. As developers, ensuring seamless, intuitive, and robust interactions is paramount. Let's delve into the best practices for these XR interaction methods by starting with VR controller input configurations.

VR controller input configuration

In the dynamic world of VR, the way the user interacts with the environment can significantly shape their overall experience. VR controllers are central to this interaction, and there are general conventions that many developers follow to ensure a consistent and intuitive user experience. The following is a look at common VR controller inputs and their frequently associated actions:

VR controller input	Primary actions	Secondary actions
Trigger button	Grabbing or picking up objects	Shooting or activating a highlighted item
Grip button	Grasping or holding objects, mimicking the action of closing one's hand	Object manipulation, such as scaling or rotating
Thumbstick/touchpad	Navigation, usually through teleportation or smooth locomotion	Rotating the user's view, scrolling through menus, or adjusting settings
Face buttons (A, B, X, Y, or equivalent)	In-game actions, such as jumping, interacting, or accessing menus	Secondary modes or tools within the VR experience
Menu or system button	Accessing in-game menus, pausing, or pulling up system settings	Often a shortcut to calibrate or reset the VR view

Haptic feedback	While not a button, haptic feedback provides tactile responses to the user, signaling successful interactions, collisions, or other in-game events	

Table 9.1 – The actions associated with different VR controller inputs

In the next section, we will dive into common hand gestures in VR.

Hand gesture recognition in VR

As VR technology has evolved, hand-tracking and gesture recognition have emerged as powerful tools for a more natural and intuitive user interaction. Unlike traditional controller inputs, hand gestures utilize the nuances of human hand movement, creating a more immersive and direct interaction within the virtual environment. The following is a breakdown of commonly recognized hand gestures and their typical VR actions:

Hand gesture	Primary actions	Secondary actions
Open hand	Often signifies a relaxed state or neutral position	Hovering over objects or menu items
Closed fist	Grasping, holding, or picking up objects; mirrors the action of clenching one's hand	Can sometimes trigger a punch or hit, especially in combat-oriented experiences
Pinch (thumb and index finger)	Selecting or interacting with a specific item or UI element	Fine-tuning or adjusting objects, such as resizing or rotating
Point (extended index finger)	Pointing to or highlighting specific items or areas	Initiating teleportation or drawing attention in social VR settings
Thumbs up	Often represents affirmation, agreement, or a positive response within social VR contexts	Can sometimes trigger specific in-game actions or emotes
Palm facing outward	Stopping or blocking, particularly in narrative-driven experiences Pushing objects away or activating barriers	

Swipe (horizontal or vertical)	Navigating through menus, scrolling, or changing views	Can also be employed in certain games for actions such as slashing
Rotation (twisting hand from side to side)	Rotating objects or adjusting settings, such as volume or brightness	Might be used to change perspective or view in some applications

Table 9.2 – The actions associated with different hand gestures

To round off this topic, let's have a look at common input methods for mobile AR applications.

Mobile AR input methods

Mobile AR has opened a world of possibilities by blending the digital realm with our physical environment, all through the lens of a mobile device. Interactivity in mobile AR varies from traditional VR since it often doesn't involve specialized controllers or hand-tracking hardware. Instead, it capitalizes on the existing capabilities of smartphones and tablets. The following is a breakdown of commonly utilized mobile AR inputs and their standard actions:

Mobile AR input	Primary actions	Secondary actions
Touch (single tap)	Selecting or interacting with an AR object or UI element	Triggering animations, accessing information, or playing videos
Touch (double tap)	Resetting or recentering an AR experience	Toggling between AR modes or views
Touch (long press)	Initiating a drag or move function for AR objects	Can also activate context-specific menus or options
Pinch and zoom	Scaling AR objects, making them larger or smaller	Zooming into detailed information or images
Swipe	Rotating AR objects or navigating through AR menus or slides	Dismissing AR elements or accessing additional content
Device movement (tilting or panning)	Exploring the AR environment, changing the view, or influencing AR object behavior	Can also be employed in games or experiences to control avatars or vehicles
Camera feed	Scanning or recognizing markers, patterns, or objects to initiate AR overlays	Capturing images or videos with AR elements superimposed

Voice commands	Controlling the AR experience or interacting with AR objects through speech	Searching for information, initiating calls, or accessing device functions
Gyroscope and accelerometer	Detecting device orientation and movement, which can influence the AR content's positioning and behavior	Used in games or simulations to steer, navigate, or control AR elements based on device tilt

Table 9.3 – The actions associated with different mobile AR inputs

Having delved into the best practices in the XR space, the next section of this chapter explores some useful toolkits and plugins that you might want to consider for your upcoming XR endeavors.

Other useful toolkits and plugins for XR development

After learning about the trends and best practices of XR development in this chapter, you're likely already formulating an idea about which area of XR development you're passionate about. As you venture further into your chosen specialization, this section will guide you by offering a detailed look at valuable XR toolkits and plugins for Unity. While we've already discussed the XR Interaction Toolkit, AR Foundation, ARKit, and ARCore, there's a plethora of other tools out there. Some of these toolkits are tailored for specific projects, while others might become staple tools for a wide range of your XR endeavors, depending on the direction you take.

Exploring powerful AR toolkits from the Unity Asset Store

Within the expansive **Unity Asset Store**, there exists a multitude of AR toolkits, each bringing its own unique capabilities to the table. This chapter will introduce you to four powerful AR toolkits and help you discern the best fit for your project needs. One of these toolkits is the **Vuforia Engine**.

Vuforia Engine, the AR pioneer

Vuforia Engine stands as one of the most widely used AR platforms globally, renowned for its robust capabilities in developing cross-platform AR applications. It targets both handheld devices and digital eyewear and offers advanced computer vision capabilities, including recognition of images, objects, and spaces.

Powering over 80,000 apps, Vuforia Engine boasts a clientele that includes big names such as *EA*, *Square Enix*, *LEGO*, and *Activision*. If giants in the industry are putting their trust in Vuforia, there's a compelling reason to consider it for your AR endeavors. If you are looking to find a job in the field of AR, gaining experience with the Vuforia Engine should be at the top of your list once you start working on further AR projects beyond this book. From adding Vuforia Engine to your Unity project and setting up **Image Targets** to testing your AR app, every step is detailed in its documentation (`https://library.vuforia.com/getting-started/getting-started-vuforia-engine-unity`).

You can find the toolkit by searching for it on the Unity Asset Store or by clicking on the following link: `https://assetstore.unity.com/packages/templates/packs/vuforia-engine-163598`

Another non-negotiable toolkit for anyone who's serious about AR development is **AR Foundation Remote 2.0**, which facilitates testing AR applications on smartphones a lot. You will learn more about it in the next section.

AR Foundation Remote 2.0 toolkit, the AR development life cycle companion

Alongside the Vuforia Engine, the AR Foundation Remote 2.0 toolkit is one of those tools you should familiarize yourself with if you want to venture further into the world of AR development. This tool will dramatically transform your AR development journey as it allows for rapid iteration and an efficient workflow.

Traditional AR development involves making changes, building the project, deploying it to a device, and then testing it. This cycle, while essential, is time-consuming. With AR Foundation Remote 2.0, developers can run and debug AR apps directly within the Unity Editor. This drastically cuts down development time, enabling more focus on refining and improving the AR experience.

The toolkit goes beyond the basic `Debug.Log()` method by offering real-time debugging capabilities. One of the challenges of AR development is visualizing how an app looks and feels on a real device. AR Foundation Remote 2.0 streams video from the Unity Editor to a real AR device such as a smartphone, letting you preview your work without a full build. This real-time feedback is invaluable for fine-tuning the user experience. With full access to the scene hierarchy and all object properties right in the Unity Editor, the debugging process becomes more intuitive and comprehensive.

Another notable feature of this tool is its **Input Remoting** feature, which lets you stream multi-touch inputs from an AR device such as a smartphone or even simulate touch using a mouse within the Unity Editor. This flexibility makes testing and tweaking user interactions more thorough and efficient.

Written in pure C# without third-party libraries, and with full source code available, this toolkit offers developers the freedom to tweak, modify, and customize according to their specific needs. It also offers seamless integration with ARKit and ARCore.

Given all of these and many more features that this toolkit offers, we highly recommend you invest in this asset to continue diving into the world of AR development. By slashing the development and debugging time, the toolkit will provide a tangible return on investment for you. Your initial cost is quickly offset by the saved hours and the elevated quality of your AR apps, and by making you a more specialized AR developer.

You can find the toolkit by searching for `AR Foundation Remote 2.0` in the Unity Asset Store or by clicking on this link: `https://assetstore.unity.com/packages/tools/utilities/ar-foundation-remote-2-0-201106`

Now that you have gained insights into which toolkits will elevate your AR development capabilities, let's have a look at some more specialized toolkits that will help you to develop specific AR applications more quickly and efficiently.

GO Map – 3D Map for AR Gaming, a hidden gem for location-based AR apps

Whenever you are developing a location-based AR game or require a rich selection of map visualization styles, such as terrains, satellite imagery, and hybrid maps, the **GO Map – 3D Map for AR Gaming** asset from the Unity Asset Store will be invaluable to you. It is also loaded with demo scenes showcasing a myriad of styles – from classic flat maps to hybrid maps, and from real building renderings to terrains.

The GO Map package allows a seamless integration of points of interest and real-world landscapes into the gameplay. As every setting and graphic control can be manipulated directly from the **Inspector** window, you can save precious development time, allowing you to focus on the content and gameplay.

You can find the toolkit by visiting the Unity Asset Store and searching for `GO Map - 3D Map for AR Gaming` or by directly navigating to its URL (`https://assetstore.unity.com/packages/tools/integration/go-map-3d-map-for-ar-gaming-68889`).

If you are more interested in real-world AR navigation or tours instead of 3D maps, the next toolkit is for you.

AR+GPS Location, enabling navigation, tours, and location-based games of the future

If games such as Pokémon GO and features such as Google Maps' AR navigation excite you, Unity's **AR+GPS Location** package will be a very interesting asset for you, as it allows you to create similar applications of your own in a relatively short amount of time.

At the forefront of this asset's offering is its AR navigation system. Drawing its strength from the *Mapbox Directions API*, the system allows developers to not just create, but also visualize routes within real-world landscapes. Whether it's a path charted by Mapbox or a personalized route crafted for lesser-known terrains, the user's journey is augmented with detailed signposts, arrows, and visual cues. A complementary feature provides a simultaneous 2D map view alongside the AR display using the integrated Mapbox SDK, offering an enriched navigational experience.

With the asset, the world truly becomes your canvas. Developers can anchor 3D objects within specific geographical locations. Whether it's a monument in a bustling city or an artifact in a remote village, these objects can be virtually situated with latitude, longitude, and altitude coordinates, breathing life into the surroundings.

Further enhancing user engagement are the AR hotspots. These are specialized zones that, when entered or approached, activate specific AR manifestations. Imagine walking through a historic alley and having a virtual guide recount its rich past, all triggered by your mere presence in that location.

Textual representations get a makeover with the asset's ability to establish 3D text markers over real-world landmarks. A tourist navigating a city could instantly get information about a site, with hovering textual cues guiding their exploration.

As the user moves, the AR overlays respond naturally and with precision, ensuring a harmonious blend between the user's movement and the augmented display. Additionally, objects can be orchestrated to move or remain stationary along intricate paths.

Adopting the AR+GPS Location asset necessitates a certain alignment of tools and hardware. Compatibility with AR Foundation versions 4.x or 5.x, or Vuforia version 10 or newer, is essential. Devices must either support ARKit for iOS or ARCore for Android when deploying with AR Foundation. For those working with Vuforia, a device with ground plane capabilities is mandatory.

You can find this asset in the Unity Asset Store by searching for `AR + GPS Location` or by clicking this link: `https://assetstore.unity.com/packages/tools/integration/ar-gps-location-134882`

By exploring the AR toolkits mentioned in this section, you are on your way to mastering AR development. The next section introduces you to similarly powerful VR toolkits that you should keep an eye on if you want to venture further into the world of VR development.

Exploring powerful VR toolkits from the Unity Asset Store

The XR Interaction Toolkit is undeniably a cornerstone for VR development in Unity and should be the tool you focus most of your time and energy on. However, there are several other toolkits that can complement or even substitute it in certain contexts. These alternatives often provide specialized features not currently found in the XR Interaction Toolkit. Diving into these toolkits can not only enhance your VR development expertise, making you a more versatile VR developer, but it can also optimize your workflow based on each project's requirements. Although the XR Interaction Toolkit is continuously evolving, getting acquainted with other hardware- or software-specific toolkits and plugins remains a crucial step for any committed VR developer, as you will learn in the next section.

Oculus Integration, SteamVR, VIVE Input Utility, and similar VR toolkits

Delving into the XR Interaction Toolkit is a great starting point, but to truly advance your VR development prowess, it's beneficial to explore the toolkits and plugins offered by VR hardware manufacturers directly. This encompasses the Oculus Integration package tailored for Meta Quest series devices, the PICO Unity Integration SDK for PICO headsets, the SteamVR plugin compatible with a broad range of VR devices, and the VIVE Input Utility asset, suitable for various headsets from VIVE and other brands. Let's explore each of these in depth to understand their significance in your journey to further refine your VR expertise:

- **Oculus Integration package**: This package provides comprehensive support for Oculus VR devices and some devices compatible with OpenVR. It enables seamless handling of all in-app audio and sound effects, provides the tools to incorporate Oculus Avatars into apps, facilitates the synchronization of avatar lip movements with speech to enhance realism, allows the inclusion of Oculus platform solutions in apps to broaden their functionality, and introduces immersive voice interactions to let players interact with the virtual world using their voice, among many other powerful features. As you can see, diving into this toolkit is pivotal for XR developers targeting Oculus devices, as it provides a comprehensive suite of tools that ensures apps are optimized for this platform (`https://assetstore.unity.com/packages/tools/integration/oculus-integration-82022`).

- **PICO Unity Integration SDK**: This is a dedicated Unity SDK crafted by PICO. It offers a wide array of features, ranging from rendering and tracking to MR features. The SDK comes packed with resources tailored for spatial audio, **eye-tracked foveated rendering**, which optimizes rendering based on the eye's gaze, **fixed foveated rendering**, which is a static gaze-centric rendering method to decrease the resolution from center to edge, eye, face, hand, and body tracking, the ability to capture mixed reality videos, passthrough, and many other features. Similar to the Oculus Integration package, this SDK is an all-inclusive package for developers aiming to create XR experiences tailored to the PICO ecosystem (`https://developer-global.pico-interactive.com/resources/#sdk`).

- **SteamVR plugin**: Maintained by Valve, this plugin is a universal interface compatible with a variety of PC VR headsets. It streamlines the representation of VR controllers in the virtual space, offers skeletal hand data, and enables intricate hand-object interactions. While this plugin has many intersections with the XR Interaction Toolkit, VR developers should feel comfortable with it too, as the toolkit or parts of its scripts are still used by various companies (`https://assetstore.unity.com/packages/tools/integration/steamvr-plugin-32647`).

- **VIVE Input Utility**: This is a versatile toolkit especially oriented toward VIVE devices but it also supports a multitude of platforms. It empowers developers to create immersive movement and object interactions. It also facilitates specific device-role associations, promotes standardized VR development, and provides cross-platform hand-tracking. For those XR developers eyeing the VIVE ecosystem, VIVE Input Utility is an essential toolkit (`https://assetstore.unity.com/packages/tools/integration/vive-input-utility-64219`).

There are many other interesting VR plugins provided by each of these manufacturers and companies. We advise you to explore each company's Unity Asset Store page to discover additional VR plugins that may spark your interest:

- Meta Quest: `https://assetstore.unity.com/publishers/25353`
- Valve: `https://assetstore.unity.com/publishers/12026`
- HTC Vive: `https://assetstore.unity.com/publishers/21581`

If you're keen on delving further into VR game development, the toolkit presented in the next section will certainly interest you.

Hurricane VR, an XR Interaction Toolkit alternative for VR game development

While the XR Interaction Toolkit is the leading toolkit for handling interactions in XR, its generalist approach can sometimes fall short for specific VR gaming applications. If you want to venture into the avenue of VR game development, **Hurricane VR** could be an interesting alternative for you as it leans heavily into the gaming domain.

This toolkit offers a rich palette of features tailored for VR games such as weapon systems, specialized physics interactions such as flipping a knife in your hand, or intricate player controllers. While the XR Interaction Toolkit provides a wide array of movements, Hurricane VR offers more nuanced player controller features for VR gaming, such as seamlessly switching between sitting and standing modes, sprinting, crouching, and many more.

The toolkit isn't just a toolbox; it's an entire workshop. With diverse samples ranging from physics interactables (such as doors and dials) to over-the-shoulder backpack inventories and weapons, developers get a hands-on experience of the toolkit's potential. Moreover, it seamlessly integrates with other assets such as *VR - Physics Interactions Bundle* and *Final IK*, amplifying its utility. You can find the toolkit when searching for it via the Unity Asset Store or directly navigating to `https://assetstore.unity.com/packages/tools/physics/hurricane-vr-physics-interaction-toolkit-177300`.

Having delved deep into toolkits that enhance your XR capabilities, it's now time to refine your XR development process. In the upcoming section, you'll discover how to elevate your XR development efficiency by leveraging the strengths of AI.

Exploring powerful AI toolkits from the Unity Asset Store

Regardless of your specific interest within the realm of XR, it's essential to utilize cutting-edge tools throughout the development life cycle of your XR projects. Leveraging AI allows you to concentrate on the core game mechanics and logic of your XR application, eliminating the need to spend countless hours on basic scripting. To be a proficient XR developer who remains updated with contemporary trends and advancements, we advise you to explore the AI toolkit outlined in the following section.

AI Toolbox for ChatGPT and DALL·E, your companion for XR development

The interplay of game development and AI has never been so seamless or potent. The **AI Toolbox for ChatGPT and DALL·E** is an embodiment of this evolution, presenting a paradigm shift in how developers perceive, interact with, and utilize code in their projects.

Imagine, instead of manually crafting each line of your C# script or shader, you simply write your requirements in human language. **ChatGPT** translates your requirements into functional and contextually relevant code, directly inside Unity. By using this toolkit, seasoned developers can streamline their workflows, dedicating more time to the heart and soul of game design.

The power of this toolkit doesn't end at scripting. **DALL·E**, a cutting-edge diffusion model, can easily transform your text descriptions into vivid images inside the Unity Engine as well. Whether it's a unique texture for your terrains or a defining logo for your game's brand, it's all achievable via a simple description.

With this asset, the Unity environment remains unaltered. Just replace your traditional **Add Script** button with the new **Generate Script** option inside the Unity Editor after installing the toolkit, and watch ChatGPT in action.

While this toolbox revolutionizes the XR development process, it's pivotal to understand its boundaries. The AI, remarkable as it is, may not consistently achieve precision for intricate requirements. It is a complement to your developmental journey, not an outright replacement.

In essence, AI Toolbox for ChatGPT and DALL·E is less a tool and more a partner, ready to share the load and elevate your game development pursuits to unparalleled heights. You can find it in the Unity Asset Store by searching for `AI Toolbox for ChatGPT and DALL·E` or by directly navigating to its URL (`https://assetstore.unity.com/packages/tools/ai-ml-integration/ai-toolbox-for-chatgpt-and-dall-e-250892`).

> **Important note**
>
> At the time of writing this book, this asset provides experimental support for Google Bard. As you are reading this, the asset might fully support Google Bard and hence might be renamed slightly. If you have trouble finding this asset in the Unity Asset Store, search for the asset's publisher, Dustyroom (`https://assetstore.unity.com/publishers/16150`), in the Unity Asset Store and check out its assets to find the potentially renamed asset.

In the following section, you will discover another powerful AI tool for creating characters in your XR scene.

Blaze AI Engine, a powerhouse for universal AI character creation

Whether you are sketching out a simple cube that metamorphoses into a vigilant patrolling agent or sculpting a tactically advanced enemy that adapts to the game environment, the **Blaze AI Engine** toolkit allows you to add these characters into your project with ease. It infuses any GameObject with intelligent and realistic behaviors without requiring you to write a single line of code.

The beauty of this asset is that you don't have to abide by a strict framework. You're free to sculpt AI behaviors as you envision them. This asset is compatible with virtually any system or asset, including visual scripting.

This toolkit isn't only about intelligent enemy creation. It also enables you to craft loyal companions to accompany players through their quests, responding dynamically to a variety of commands.

To enrich your future XR endeavors with an endless range of AI characters, check out the asset in the Unity Asset Store (`https://assetstore.unity.com/packages/tools/behavior-ai/blaze-ai-engine-194525`) by searching for `Blaze AI Engine`.

The next section introduces you to a useful tool that simplifies prototyping your XR applications.

Prototyping XR applications

ShapesXR (`https://www.shapesxr.com`) offers a range of features for UI/UX and spatial model prototyping. It simplifies 3D creation, allowing users to start with basic elements or choose from an extensive library of primitives. With the ability to customize colors, materials, and text, the platform offers precise control through its snapping system.

Collaboration is also a central focus of ShapesXR. It facilitates real-time co-creation by enabling team members to enter VR environments easily. This immersive approach to designing reviews can enhance understanding and productivity.

ShapesXR supports various 2D and 3D formats, simplifying the integration of assets. A *Figma* plugin keeps assets synchronized, and you can import and export assets seamlessly from your browser.

ShapesXR also offers MR capabilities, enabling designs to be brought into the real world. Features such as passthrough materials provide valuable insights for XR development.

Summary

Reflecting on your journey through this book, it began with you mastering the foundational elements of Unity, progressed to the creation of your first VR and AR applications, and culminated in the development of an immersive drumming experience, experimenting with eye-tracking, multiplayer functionalities, and sound and visual effects. With this chapter's close, you should not only possess the confidence to design, develop, and launch intermediate XR applications but also be primed to delve deeper into XR development. Armed with thoughts of toolkits you're eager to discover next and comprehensive guidance on structuring your XR initiatives, you're now poised to not just create XR solutions on your own, but also oversee or guide them.

Index

S

T

‹packt›

www.packtpub.com

Subscribe to our online digital library for full access to over 7,000 books and videos, as well as industry leading tools to help you plan your personal development and advance your career. For more information, please visit our website.

Why subscribe?

- Spend less time learning and more time coding with practical eBooks and Videos from over 4,000 industry professionals
- Improve your learning with Skill Plans built especially for you
- Get a free eBook or video every month
- Fully searchable for easy access to vital information
- Copy and paste, print, and bookmark content

Did you know that Packt offers eBook versions of every book published, with PDF and ePub files available? You can upgrade to the eBook version at packtpub.com and as a print book customer, you are entitled to a discount on the eBook copy. Get in touch with us at customercare@packtpub.com for more details.

At www.packtpub.com, you can also read a collection of free technical articles, sign up for a range of free newsletters, and receive exclusive discounts and offers on Packt books and eBooks.

Other Books You May Enjoy

If you enjoyed this book, you may be interested in these other books by Packt:

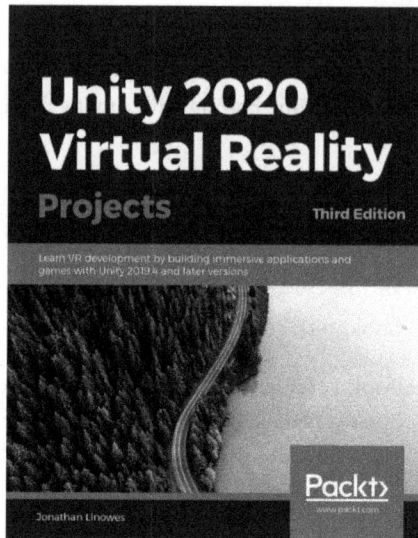

Unity 2020 Virtual Reality Projects

Jonathan Linowes

ISBN: 978-1-83921-733-3

- Understand the current state of virtual reality and VR consumer products
- Get started with Unity by building a simple diorama scene using Unity Editor and imported assets
- Configure your Unity VR projects to run on VR platforms such as Oculus, SteamVR, and Windows immersive MR
- Design and build a VR storytelling animation with a soundtrack and timelines
- Implement an audio fireball game using game physics and particle systems
- Use various software patterns to design Unity events and interactable components
- Discover best practices for lighting, rendering, and post-processing

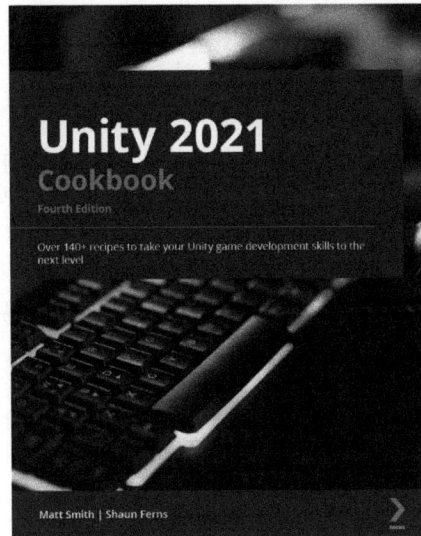

Unity 2021 Cookbook

Matt Smith, Shaun Ferns

ISBN: 978-1-83921-761-6

- Discover how to add core game features to your projects with C# scripting
- Create powerful and stylish UI with Unity's UI system, including power bars, radars, and button-driven scene changes
- Work with essential audio features, including background music and sound effects
- Discover Cinemachine in Unity to intelligently control camera movements
- Add visual effects such as smoke and explosions by creating and customizing particle systems
- Understand how to build your own Shaders with the Shader Graph tool

Packt is searching for authors like you

If you're interested in becoming an author for Packt, please visit `authors.packtpub.com` and apply today. We have worked with thousands of developers and tech professionals, just like you, to help them share their insight with the global tech community. You can make a general application, apply for a specific hot topic that we are recruiting an author for, or submit your own idea.

Share Your Thoughts

Now you've finished *XR Development with Unity*, we'd love to hear your thoughts! Scan the QR code below to go straight to the Amazon review page for this book and share your feedback or leave a review on the site that you purchased it from.

`https://packt.link/r/1-805-12812-4`

Your review is important to us and the tech community and will help us make sure we're delivering excellent quality content.

Download a free PDF copy of this book

Thanks for purchasing this book!

Do you like to read on the go but are unable to carry your print books everywhere?

Is your eBook purchase not compatible with the device of your choice?

Don't worry, now with every Packt book you get a DRM-free PDF version of that book at no cost.

Read anywhere, any place, on any device. Search, copy, and paste code from your favorite technical books directly into your application.

The perks don't stop there, you can get exclusive access to discounts, newsletters, and great free content in your inbox daily

Follow these simple steps to get the benefits:

1. Scan the QR code or visit the link below

https://packt.link/free-ebook/9781805128120

2. Submit your proof of purchase
3. That's it! We'll send your free PDF and other benefits to your email directly

www.ingramcontent.com/pod-product-compliance
Lightning Source LLC
Chambersburg PA
CBHW061806210326

41599CB00034B/6897